A Ditch in Time

The City, the West, and Water

Patricia Nelson Limerick
with Jason L. Hanson

A Ditch in Time

The City, the West, and Water

Patricia Nelson Limerick

with Jason L. Hanson

and the assistance of Timothy Brown, Dylan Eiler, Christian Heimburger, Buzzy Jackson, and Alex Lande

FULCRUM
GOLDEN, COLORADO

Authors' Note: Denver Water provided funding to support some of the background research for this book. By agreement between the authors and the agency, this funding carried no strings and placed no constraints on the book's content, analysis, and interpretation. The authors alone hold responsibility for the book's style, tone, perspective, and findings.

Library of Congress Cataloging-in-Publication Data
Limerick, Patricia Nelson, 1951-
A ditch in time : the city, the west, and water / Patricia Nelson Limerick with Jason L. Hanson.
p. cm.
Includes bibliographical references and index.
ISBN 978-1-55591-366-3 (pbk.)
1. Denver (Colo.). Water Dept.--History. 2. Water resources development--Colorado--Denver--History. I. Hanson, Jason L. II. Title.
TC425.C6L56 2012
363.6'10978883--dc23

2012012386

Printed in the United States of America
0 9 8 7 6 5 4 3 2 1

Design by Jack Lenzo

Fulcrum Publishing
4690 Table Mountain Dr., Ste. 100
Golden, CO 80403
800-992-2908 • 303-277-1623
www.fulcrumbooks.com

To Houston Kempton,
My companion in the flow of time

Contents

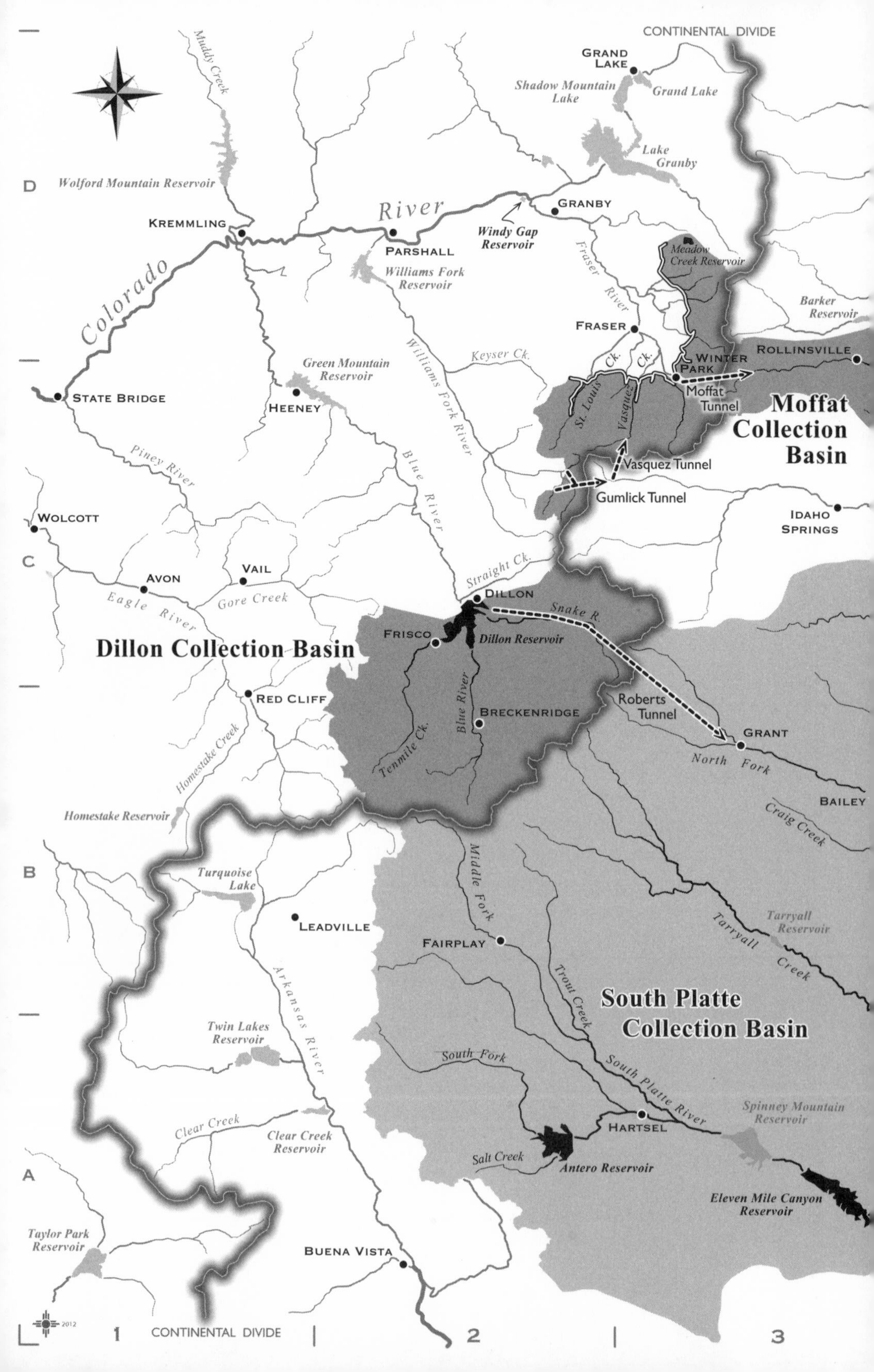

CONTINENTAL DIVIDE
GRAND LAKE
Shadow Mountain Lake
Grand Lake
Lake Granby
Muddy Creek
Wolford Mountain Reservoir
D
KREMMLING
Colorado River
PARSHALL
GRANBY
Windy Gap Reservoir
Williams Fork Reservoir
Fraser River
Meadow Creek Reservoir
Barker Reservoir
FRASER
Keyser Ck.
Williams Fork River
Green Mountain Reservoir
WINTER PARK
ROLLINSVILLE
St. Louis Ck.
Vasquez Ck.
Moffat Tunnel
Moffat Collection Basin
STATE BRIDGE
HEENEY
Blue River
Piney River
Vasquez Tunnel
Gumlick Tunnel
IDAHO SPRINGS
WOLCOTT
C
AVON
VAIL
Straight Ck.
DILLON
Eagle River
Gore Creek
Snake R.
FRISCO
Dillon Reservoir
Dillon Collection Basin
RED CLIFF
BRECKENRIDGE
Roberts Tunnel
GRANT
Tenmile Ck.
Homestake Creek
North Fork
BAILEY
Homestake Reservoir
Craig Creek
B
Turquoise Lake
Middle Fork
LEADVILLE
Tarryall Reservoir
Tarryall Creek
FAIRPLAY
Trout Creek
Arkansas River
South Platte Collection Basin
Twin Lakes Reservoir
South Fork
South Platte River
Spinney Mountain Reservoir
Clear Creek
Clear Creek Reservoir
HARTSEL
Salt Creek
Antero Reservoir
A
Eleven Mile Canyon Reservoir
Taylor Park Reservoir
BUENA VISTA
2012
1
CONTINENTAL DIVIDE
2
3

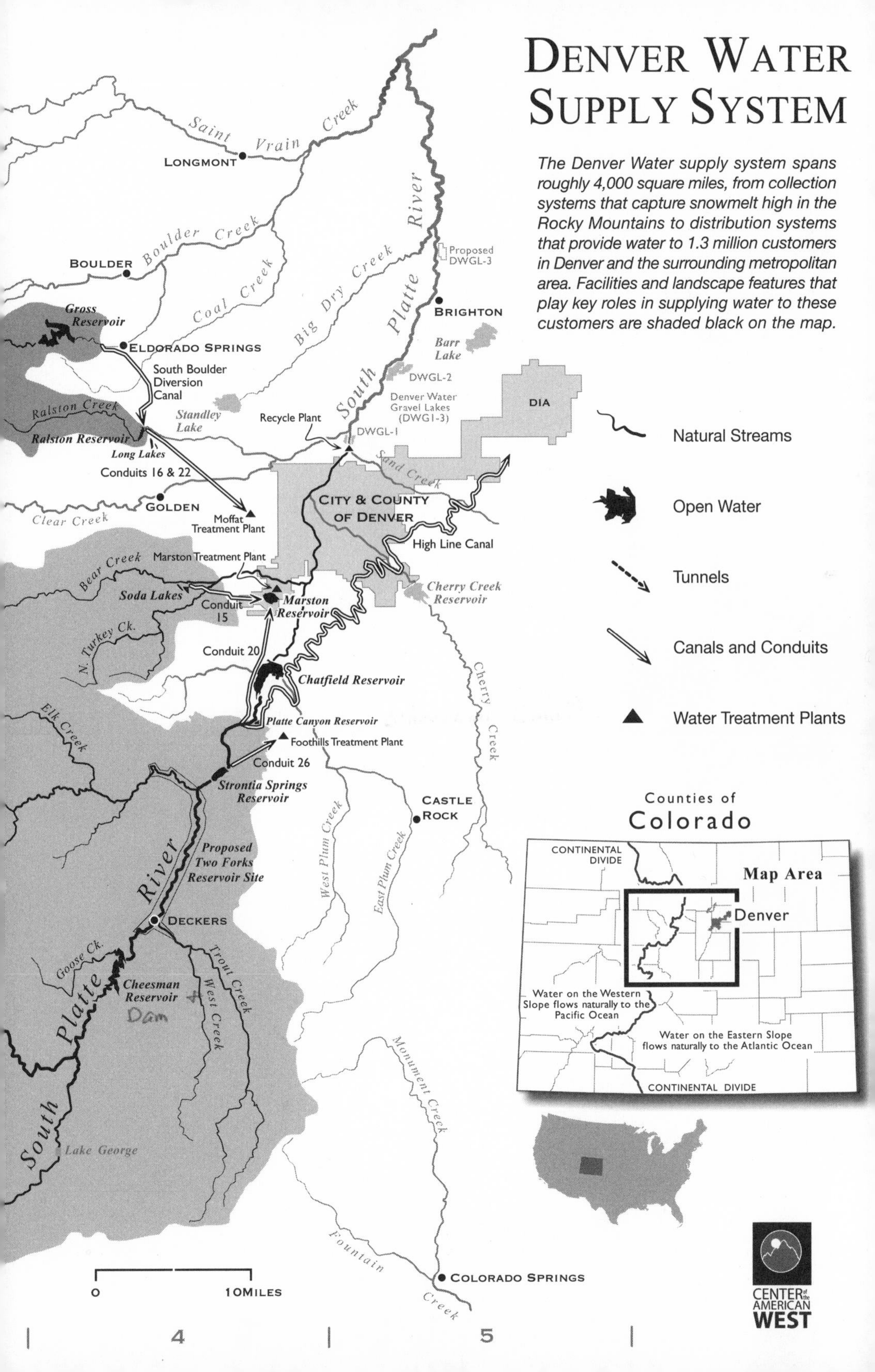
Denver Water Supply System
The Denver Water supply system spans roughly 4,000 square miles, from collection systems that capture snowmelt high in the Rocky Mountains to distribution systems that provide water to 1.3 million customers in Denver and the surrounding metropolitan area. Facilities and landscape features that play key roles in supplying water to these customers are shaded black on the map.
Natural Streams
Open Water
Tunnels
Canals and Conduits
Water Treatment Plants
Saint Vrain Creek
Longmont
Boulder
Boulder Creek
Coal Creek
Big Dry Creek
South Platte River
Proposed DWGL-3
Brighton
Gross Reservoir
Eldorado Springs
South Boulder Diversion Canal
Barr Lake
DWGL-2
Denver Water Gravel Lakes (DWG1-3)
DIA
Ralston Creek
Ralston Reservoir
Standley Lake
Recycle Plant
DWGL-1
Long Lakes
Conduits 16 & 22
Sand Creek
Golden
Clear Creek
Moffat Treatment Plant
City & County of Denver
High Line Canal
Bear Creek
Marston Treatment Plant
Soda Lakes
Conduit 15
Marston Reservoir
Cherry Creek Reservoir
N. Turkey Ck.
Conduit 20
Chatfield Reservoir
Cherry Creek
Elk Creek
Platte Canyon Reservoir
Foothills Treatment Plant
Conduit 26
Strontia Springs Reservoir
Castle Rock
Proposed Two Forks Reservoir Site
West Plum Creek
East Plum Creek
Deckers
Goose Ck.
Cheesman Reservoir
Dam
Trout Creek
West Creek
Monument Creek
Lake George
Fountain Creek
Colorado Springs
0
10Miles
Counties of Colorado
Continental Divide
Map Area
Denver
Water on the Western Slope flows naturally to the Pacific Ocean
Water on the Eastern Slope flows naturally to the Atlantic Ocean
Center of the American West

Major Reservoirs of the Denver Water System

Reservoir	Capacity (Acre-Feet)	System Function
Antero	19,881	Drought Supply
Chatfield	27,428	Replacement/Drought Supply
Cheesman	79,064	Seasonal/Drought Supply
Dillon	257,304	Seasonal/Drought Supply
Eleven Mile	97,779	Drought Supply
Gross	41,811	Seasonal/Drought Supply
Long Lakes	1,787	Seasonal
Marston	19,796	Terminal Storage
Meadow Creek	5,370	Seasonal Supply
Platte Canyon	910	Seasonal Supply
Ralston	10,776	Terminal Storage
Soda Lakes	645	Seasonal Supply
Strontia Springs	7,863	Regulating/Terminal Storage
Williams Fork	96,822	Replacement/Substitution
Wolford Mountain	25,610	Substitution

Acknowledgments

This book took a long time to come into being.

My own history as a writer figures in the explanation of its long journey to print. Exactly twenty-five years ago, my second book—*The Legacy of Conquest: The Unbroken Past of the American West*—was published. This book has had an adventure-packed life of its own. Denounced by some and welcomed by many, *Legacy* surprised its author and publisher by becoming a "classic": required reading for many undergraduate and graduate students, but also widely read by many in the general public.

If I had been ambitious and hungry for professional achievement before 1987, contentment over the happy destiny of *The Legacy of Conquest* seemed to put me out of business when it came to the writing of book-length manuscripts. I seemed to have made a permanent switch of genres. To the understandable dismay of editors who had, in good faith, added their signature to mine on book contracts, I churned out essays and articles, as well as the less-exalted literary expressions called reports and memos. More than anything, I became a one-woman, 24/7 production facility for speeches, talks, and lectures.

Was I freaked out by the success of *Legacy* and thereby unwilling to run the risk of writing and publishing another book? While the pleasant burden of that book's success might have played a part in my choices, a greater factor was the founding of the Center of the American West at the University of Colorado with my partner, law professor Charles Wilkinson. The robust health of this organization unleashed a wild round of activities perfectly suited for a person grateful to have been born and raised before the diagnosis of attention deficit disorder came into play.

And yet, keeping the Center of the American West in financial well-being meant staying on the lookout for funding opportunities. Various figures at the Denver Water Department held the convincing opinion that their agency had an instructive and important history. Soon after I moved to Colorado, in 1984, I had come to know Chips Barry, who became manager of Denver Water in 1991. Chips, as readers of this book and especially its afterword will know, was a remarkable person. His respect for history matched, in intensity, his sense of humor. An agreement between Denver Water and the Center of the American West, reserving full intellectual independence for the center, launched this project.

In the early years, the plan was not a strenuous one. We would employ graduate students and postdoctoral affiliates to write chapters for a book that would have, as its principal asset and charm, a robust collection of photographs from the Denver Water archives. We began with the assumption, now hard to reconstruct, that the written text would be lite and not particularly weighted with thought, reflection, and interpretation.

The result was not spirit lifting.

I finally settled down and shouldered the burden that should have been mine all along. I took on the job of fundamentally reworking the cobbled-together manuscript. There was some comedy in this situation, since my working circumstances were aided by a parallel and more literal project in remodeling. In January 2010, I moved into my husband's house so that we could remodel my house. I took only books, articles, and notes related to *A Ditch In Time.* This put a valuable damper on my ordinarily impressive gift for conjuring up distractions.

As the contractor tore apart my house, built new rooms, and reconfigured some of the preexisting structure, I performed a comparable set of actions with the manuscript that, for a spell, ruled my life. By October 2011, I had created a draft that seemed robust, unified, funny, and grounded in the two fields of western history and water policy.

This brings me to the acknowledgments.

On the title page, the author identification appears with these words: "Patricia Nelson Limerick with Jason L. Hanson," followed by a phalanx of collaborators and assistants. Christian Heimburger launched the research process years ago. In a game effort, Tim Brown and Buzzy Jackson wrote chapters for the first draft, which bears little resemblance to the current text. (Both Tim and Buzzy, it is important to say, are gifted and original writers when they are set free of the production-by-committee method under which they were forced to operate!) Two very talented recent graduates of the University of Colorado—Dylan Eiler and Alex Lande—did further research and extensive fact-checking.

Nearly all the work of the Center of the American West is collaborative and cooperative, closer to the customs of a team of scientists than the more individualistic ways of humanists. This leaves us in a state of some perplexity when it comes to the usual allocations of intellectual property. I wrote the text in its current version, and readers familiar with my other work will consistently recognize my voice and style. But Jason Hanson played a crucial role in the book's completion, and his name therefore appears with mine on the title page. Jason coordinated the follow-up research when I identified new topics we had to add, and he fact-checked and proofread within an inch of his life (or what would have been within an inch of his life if he were not a person of such unusual vitality). He wrangled the photographs, acquired the permission to use them, and wrote the captions. He did all this with world-class equanimity and good nature.

It is impossible to imagine how I and my comrades could have powered through to the completion of this manuscript without the gifts and insights of our captain at the center, Kurt Gutjahr. His official title is program director, although plenty of other terms apply equally well: executive director, managing director, coordinator, mediator, problem solver, navigator, conductor, therapist, and coach. Kurt is the gyroscope of the center of the American West. He gets the credit for a great share of the center's success.

Many others who worked or work at the center have contributed to the completion of this book: Roni Ires, Amanda Hardman, Ryan Rebhan, Jennifer Aglio, Claudia Puska, Adrianne Kroepsch, Raissa Johnson, Sam Chapman, and Ashley Howe have all pushed this book forward with their hard work and lively spirit. For this project and quite a number of others, Honey Lindburg deserves a special note of thanks. Purchasers of *A Ditch in Time*, as well as recipients of many Center of the American West announcements and documents, are beneficiaries of the fact that Honey, a gifted graphic designer and visual artist, has expertly taken on the challenge of working in an organization led by an excessively word-oriented person.

Another crucial member of the team is cartographer Jim Robb. His map of the Denver Water supply system appears in the front of the book. Readers are encouraged to consult this map often, since words can only go so far in conveying the complicated spatial and geographical terms of this vast array of streams, rivers, dams, reservoirs, tunnels, ditches, and treatment plants.

Without the opportunity to consult noted hydrologist Dan Luecke, this project would have been far more difficult. A scientist capable of crystal-clear communication of his expertise, Dan is also a historic figure in this story. Similarly, Eric Kuhn at the Colorado River Water Conservation District and Eric Wilkinson at the Northern Colorado Water Conservancy District have played and continue to play a key role in Denver Water's unfolding history, and their perspectives enriched our understanding of the story we have told here. I am grateful, as well, to a number of other friends with deep experience in the area of natural resources management who read the manuscript and provided thoughtful responses: Ruth and Ken Wright, Dave and Jan Robertson, Pam and George Beardsley, and Ron Stewart. Ann Heinz, the dean of Continuing Education at the University of Colorado, read the full manuscript and provided both encouragement and a number of useful corrections. Mark Shively and George Sibley read parts of the manuscript and, true to their characters, provided thought-provoking and well-informed commentary. In an early stage of the project, my wonderful friend Sue Deans pitched in to the cause of helping us sort out tedious detail from necessary and telling fact. Two very accomplished environmental historians, Martin Melosi and Martin Reuss, read the manuscript closely and offered comments and suggestions that helped me tremendously. Sunnie Bell and Adrianne Kroepsche proofread the final manuscript, and their sharp eyes enhanced the clarity and consistency of this text to the great benefit of our readers.

While scrupulously abiding by the declaration on the opening page of the author's right to exercise unrestricted judgment, various officials at Denver Water read the manuscript, called factual errors to our attention, and engaged in spirited exchanges over various aspects of their complicated work. We are grateful to manager and CEO Jim Lochhead, director of planning Dave Little, general counsel Patti Wells, and attorney Casey Funk. Before his untimely death, Chips Barry was a resourceful and valuable partner in conversation, and we are grateful to Gail Barry for the permission to include as this book's afteword the speech that he wrote for a joint presentation with me a few months before his death. Holly Geist, Laura George, and Duncan McCollum gave us invaluable guidance and help in utilizing Denver Water's rich archives. Retired manager of Denver Water community relations Jane Earle lent her support to this enterprise in a number of ways. In 2008, Jane arranged a remarkable tour of Denver Water facilities, where we had the opportunity to meet and talk with Denver Water personnel Ed Christiansen, Rusty Christensen, Mike Couts, Jade Dreier, Cindy Bryan, and Dale Beverly.

Internationally known for his photography of wildlife and wilderness, John Fielder very graciously consented to let us use his photograph of the High Line Canal for the book's cover. Two fine photographers—Jackie Shoemaker and Ted Wood—provided the contemporary images that much enrich the visual dimension of this book. Coi Drummond-Gehring guided us in accessing the expansive collection of historical photographs curated by the Denver Public Library's Western History and Genealogy Department, and Greg Moore and John Sunderland at *The Denver Post* provided the iconic photo of Chips Barry standing behind a waterwheel.

Glenn Saunders's daughter, Carol, generously and graciously lent me her father's scrapbooks. Well-stocked with the principal news coverage of Denver Water over his long career, these scrapbooks gave me the extraordinary opportunity to feel very directly in their compiler's company. On Saturdays, alone in the center offices with the record that Glenn Saunders had assembled of the agency that he served for years, I read and took notes with an unusual sense of engagement with a significant historical figure.

Colorado Supreme Court justice Gregory Hobbs is a rock star of expertise on Colorado water law. As I worked my way through the remodeling of the manuscript, Justice Hobbs read each chapter and corrected a number of flubs of interpretation and fact on my part. I am very much in his debt for his care in reviewing the manuscript, for the lively foreword that begins this book, and for the joy of being the friend of Greg and Bobbie Hobbs.

I am indebted to a number of audiences in Colorado and elsewhere who let me field-test my ideas in talks and speeches while *A Ditch in Time* was inching along toward completion. Audiences at the Center for Historical Studies at Northwestern University, the City Club of Denver, the Colorado Historical Society, the Frasier Meadows Retirement Community, the Longmont Senior Center, the Colorado chapter of the American Water Resources Association, Western State College in Gunnison's Water Workshop, and the Center for Global Humanities at the University of New England all did valuable service as my "focus groups." At the Douglas County Library in Highlands Ranch, Colorado, I benefitted both from audience commentary and from the presentation by my fellow speaker, John Hendrick, general manager of the Centennial Water and Sanitation District.

Since Robert Baron founded his publishing house in 1984, Fulcrum Publishing has taken center stage in the intellectual life of the West. Fulcrum's publications—on the region and on broader issues of nature and environment—have lit up the world, and all of us at the Center of the American West are proud to be associated with this accomplished publisher. As Fulcrum's current president and publisher, Sam Scinta has been a prince, communicating with clarity and warmth on every occasion, working out a contract by which revenue from this book goes directly to the Center of the American West, and keeping his patience with a few missed deadlines. As the book's editor, Carolyn Sobczak brought to bear exactly the fresh point of view that the manuscript needed, combining encouragement with requests for clarification in exactly the right balance.

As I have often had the occasion to remark, the position of faculty director and chair of the Board of the Center of the American West proves to be the best job on the planet. Nonetheless, I would have been happy to be spared one aspect of this job—repeatedly having to inform the enormously supportive external Board of the Center of the American West that this book was not quite finished. With that melancholy action now removed from my professional obligations, I can now simply say that I am unendingly grateful to the board members for their patience and kindness. Center board member Susan Kirk provided me with a number of useful opportunities to talk with her husband, former Denver Water Board member Dick Kirk, and by virtue of serving sequentially on the Denver Water Board and the Center of the American West Board, Hubert Farbes gave me a grounded and valuable angle on the changes of the early 1990s. Center board member John Wittemyer, after a distinguished career in water law, gave the manuscript a close reading and, when he found only a few missteps on my part, enhanced my confidence considerably.

The commitment of the Center of the American West to do this book began more than seven years ago. Over that time, I was widowed and remarried. Since we took up with each other, my second husband, Houston Kempton, has lived in the company of this seemingly endless project. It is a measure of Houston's warmth, good nature, congeniality, and endless curiosity that he has never said a word of sorrow or anger over the fact that *A Ditch in Time* has been our constant companion. My guess is, however, that we will both enjoy life more with *A Ditch in Time* out in the world on its own. The book only seemed to be chronically incomplete, but my gratitude to Houston for putting up with this undertaking is actually endless.

—Patricia Nelson Limerick
Center of the American West
University of Colorado, Boulder
March 2012

Patty covered the vast and varied terrain of our gratitude pretty thoroughly, and I have only a few words to add.

I would like to thank my colleagues at the Center of the American West over the years for pitching into this project with diverse and extraordinary talents that regularly leave me in awe. In particular, I am indebted to Kurt Gutjahr for his good cheer and steady hand in guiding the center's endeavors and to Patty for her kindness and unfailing generosity at every turn. In Patty's dedication to the West and the people who call it home, I find my inspiration.

I am also grateful beyond expression to the friends and family who have supported and encouraged me over the course of this long project. Standing foremost among this crowd is my wonderful wife, Stacie, who has been my touchstone and my greatest blessing, and our beautiful daughter, Ellery, who proves the old Denver Union Water Company claim that better water does indeed make better babies.

—Jason Hanson
Center of the American West
University of Colorado, Boulder
March 2012

Foreword

Justice Gregory J. Hobbs Jr., Colorado Supreme Court

I have walked through many lives, some of them my own, and I am not who I was, though some principle of being abides, from which I struggle not to stray.

—Stanley Kunitz, *Passing Through: The Later Poems*

The history of water development in Denver offers a particularly fine post for observing the astonishing and implausible workings of historical change and, in response, for cultivating an appropriate level of humility and modesty in our anticipations of our own unknowable future.

—Patricia Nelson Limerick, *A Ditch in Time: The City, the West, and Water*

Associate Denver Water with humility? Or, for that matter, a water historian? Here's the exception to the rule.

When the historian steps out of the archives and the classroom well-prepared to engage "at every intersection where values and meanings cross," she might just locate herself in "the thick of life," writing a credible (I might say incredible) history of a western American water utility.

I have borrowed quotations from a New York elder who lamented that twentieth-century American poets seemed to have relegated themselves to the classroom. Stanley Kunitz urged his fellow poets to catch themselves up in more "dangerous traffic, between self and universe."

Isn't Patty Limerick *the revisionist western historian*? The history of the West is the history of bad men doing bad things, the worst being imperialistic water developers. Stay with me on this.

As a result of her intensive study of Denver Water, Limerick concludes that Donald Worster and Marc Reisner failed to follow the evidence in asserting that "the history of the development of water in the West followed a plot (in both senses of the word!) that led to the centralization of power in the hands of a small, inflexible, undemocratic, entirely self-interested elite." She astonishes herself by rejecting such "well-entrenched ideas" because "until recently, I believed them myself."

Please excuse me if I call her Patty instead of Dr. Limerick. She's way overachieved this learned title through her books and innumerable public presentations throughout our great country. Her accessibility invites degrees of reinvigorating informality, accommodating our western airs. The more we feel her presence, the better situated we are in the course of coming around.

Through the Center of the American West, which she heads at the University of Colorado, she brings people together who might otherwise misunderstand their own histories and thereby prejudice others.

She's a zigzag writer, thinker, and speaker. She interseams the clay she crafts with. Just when you think you've arrived at the water hole, she's got you traversing back off some impossible pinnacle you've cragged yourself out on.

I liken her manner of relating disparate parts to the intelligible whole, the way Ancestral Puebloan people fired zigzag patterns onto everyday implements, their water jugs and mugs:

Water Pocket

There's nothing dry about Patty
despite her PhD in aridity,
she's the water pocket in the desert
for thirsty students,
the zigzag lightning stitch
on an Ancestral Puebloan
drinking mug meant to refresh
the morning star on her
journey West.

(For Patty Limerick on her sixtieth)

For her sixtieth birthday celebration, which occurred in the throes of rendering this book, she threw a dance at Boulder's Chautauqua beneath the Flatirons, where the plains meet the foothills leading up to the Great Divide. She legged up her western boots, invited a dance hall full of friends, and loosed her anxieties to a mariachi band.

After losing her beloved husband and grieving in the depths

of maybe-give-it-up, we see she's found and married an entire city: Houston. He now abides within their living room, where she conducts home chats with students, intermingling invited guests.

Thomas Jefferson never got west of the Blue Ridge Mountains. He sent others here. He thoroughly believed the perpetual future of our growing democracy would be agrarian. So did John Wesley Powell (who, despite our penchant for crediting his watershed views, predicted we must dry up all the rivers to irrigate but 5 percent of the lands for lack of enough water to grow all the farms we hope to depend upon).

Now, as Patty points out, 2 percent of us farm; the rest of us live in cities and suburban extensions. We desire to eat and play well. We depend upon municipal and special district water utilities for drinking water, showers, potties, home gardens, shade trees, and dog and volleyball lawns. We insist upon cheap, reliable service, yet blame our service providers for despoiling the rivers and draining the rural towns of their irrigated livelihoods.

Meanwhile, we have conserved considerably because we've had to. There's been a 20 percent reduction in water usage, replicated throughout the West as the early twenty-first-century drought has interrupted our preexisting habits.

The recent managers of Denver Water, Chips Barry and Jim Lochhead, like their counterparts across our treasured western landscape, are accepting the reality of climate change into their portfolio of risks and options. To all who would make this issue political, they point out limits in our abilities to predict when and where the plenty or the scarcity may occur. It's erratic.

So beware, as Patty cautions. Our prophecies and projected limitations often bewilder the would-be soothsayers. Be prepared on all fronts of conservation and supply, and cooperate with those hearty enough to remain agrarian.

Introduction

In the early nineteenth century, explorers of the Front Range of Colorado declared that the scarcity of water made conventional American settlement in that locale impossible. In the twenty-first century, the Front Range is home to a population of millions.

What happened?

An enormous infrastructure rearranged the distribution of water. Dams store spring runoff, capturing water that would otherwise flow downstream and out of the state. Tunnels under the Rockies divert water from the Western Slope to the Front Range. The comparative scarcity of the western water supply did not impose the limits that the early explorers expected because of human engineering skills and because of the creation of an enormously complicated legal and political structure. The belief that aridity in itself would exercise the power to reshape or even prohibit standard American customs of settlement and land use did not turn out to hold water. On the contrary, that power was overruled by the institutions and organizations that acquired, manipulated, and managed water.

For all the distinctive features of this particular place, the transformation of the Front Range of Colorado is one example of a much larger historical process. In a manner unparalleled in most of human history, in the United States in the last half century, millions of people lived in a condition of extraordinary material ease, supplied with an abundance of food, energy, and water by institutions and organizations to which most of the beneficiaries did not pay an ounce of attention. The degree of material good fortune achieved by Americans in this era has only been equaled by the degree of their inattention to its sources. When future historians look back at the United States in the twentieth century, if they choose accuracy over graceful phrasing, they will have reason to christen this unusual historical interlude as the Era of Improbable Comfort Made Possible by a Taken-for-Granted but Truly Astonishing Infrastructure.

This is a book about one of the organizations that built and maintained this extraordinary, comfort-supplying infrastructure. The founders and leaders of the entity we now know as Denver

Water were the opposite of procrastinators. When it came to establishing water rights and building structures to store the water thus obtained, they did not wait for crisis to push them into action. Folklore is full of parables and fables that contrast energetic and foresighted creatures who take action well ahead of need with lazy and complacent creatures who do not look ahead and who thus make it easy for trouble to catch them by surprise. In the water allocation of the American West today, the plot structure of these parables and fables appears in many areas. As the title of this book indicates,* Denver Water took action early and repeatedly to secure and provide water to the citizens in its service area, long before actual shortages could appear. In the last half of the twentieth century, as urban and suburban growth rates took off in the Front Range, other, less "proactive" communities found themselves in a pickle, with growing demands for water and a late entry into the scramble for water rights. Very much in the manner of the old fables and parables, the late arrivals looked enviously and covetously at the resources held by the foresighted and farsighted entity named Denver Water. To a remarkable degree, the prospects for future growth in the Front Range in the twenty-first century *seemed* to be under the control of Denver Water and a few counterpart organizations. The word *seemed* appears in italics because there are compelling reasons to question the extent and scale of Denver Water's imperial power over the water supply of the state of Colorado. A good share of this book conducts an exploration of those reasons.

By calling attention to the history of the Denver Water Department, I aspire to challenge the mental habit that welcomes and relishes natural resources as long as they originate in places and processes that are out of sight and out of mind. As historian Martin Melosi put it, "Service delivery is a 'hidden function' largely because it often blends so invisibly into the urban landscape." Diagnosed and lamented by many observers in recent times, this disconnection or alienation from the sites of production of energy, food, and water has proven to be an effective force for undermining and eroding a sense of responsibility. When people flipped a switch and summoned light or heat or turned a faucet and conjured, at

* For readers whose attention to platitudes may have lapsed, *A Ditch in Time* is a play on the old truism that "a stitch in time saves nine."

will, a flow of clean water, normal human curiosity—Where did this come from? Who made this happen? What consequences will come from this?—went dead. Events and trends of the early twenty-first century are, however, bringing that curiosity back to life and also reviving a sense of connection and responsibility. From the volatile price of gasoline to worries about the nation's dependence on imported oil, complacency about the use of energy has been rattled. Anxiety about afflictions carried by food, whether through bacterial contamination or chemical additives, has unsettled the capacity to take the grocery store's offerings for granted. In a similar way, public expressions of alarm over the supply and quality of water have grown in frequency and audibility. A variety of wake-up calls have interrupted the American public's long nap.[1]

When the sites of production are no longer concealed and the connection between our material comfort and a giant network of coal mines, natural gas wells, electrical generating plants, transmission lines, dams, aqueducts, trans-basin tunnels, and treatment plants stands revealed, a common first response has been not gratitude, but the condemnation of the individuals and groups who created this network and kept it in operation. Another dimension of this book is a reexamination of that understandable impulse to blame and condemn organizations like Denver Water. Experienced scholars in the field of western water history have urged readers to make fuller reckoning with the benefits and gains, as well as the losses and injuries, produced by the creation of the infrastructure that supports and supplies American communities today. Introducing his history of the Colorado–Big Thompson Project and the Northern Colorado Water Conservancy District, historian Daniel Tyler extended a useful and forceful invitation:

> With all due respect to those who view water projects as the work of evil megalomaniacs, I would ask readers to give some thought to the conditions that fostered the need for supplemental water . . . and the vision of [the] men who believed they were taking risks for the betterment of their families, friends, homes and businesses.[2]

Surveying water development on a global scale, writer Diane Ward offered a similar observation: "In times of polarized sentiment about the building of big dams, it has often been forgotten that many dams were built with the best of intentions, by men who genuinely wanted to use them for the betterment of mankind."[3]

Readers may feel an occasional temptation to condemn Denver Water's manipulation and mastery of both nature and other communities in the state. And yet there are good reasons to temper that temptation with other recognitions and other appraisals. To contemplate a structure like Cheesman Dam, and to police oneself so closely that disapproval stamps out any spark of admiration, one must undertake an intense treatment program in maintaining a properly dour and somber stance of dismay even when confronted with extraordinary human enterprise. The contemporary need for inspiration—for parables of people facing tough problems, refusing discouragement, and pressing on to solutions and remedies—is the most urgent need of the twenty-first century. Western American history presents an abundance of such case studies, but all of them come with dimensions of moral complexity and displays of the grimmer aspects of human nature. I invite readers of this book to join me in the invention and refining of methods to extract inspiration from the complicated figures of the past, while fully acknowledging their flaws, failings, and blunt exercises of force and power.

While I am not an apologist for Denver Water, working on this book has left me reluctant to offer a complacent condemnation of the organization and its leadership.

Why?

Because those of us who live in the American West today are dependent on, complicit with, and indebted to the organizations and institutions that disrupted the ecosystems and disturbed the landscapes that, a little late in the game, we came to treasure. This is a paradox that is not going to go away, and it is a source of much mischief if denied and evaded. Handled with honesty, the paradox provides the best footing we have for moving toward a more honest and productive relationship to natural resources and the managers to whom, for so long, we delegated the responsibility to acquire those resources and to supply them to us on demand.

A Ditch in Time also has the goal of acknowledging the historical and contemporary importance of cities in the American West, a region long associated with imagery of open spaces and rural enterprises. The Jeffersonian agrarian dream contributed to the desire to see the West as fundamentally rural, and the imagined West of movies and novels reinforced the preference for open spaces over urban spaces. Western historians, present company included, have displayed their own symptoms of attention deficit when it comes to reckoning with western cities. In the most influential histories of western water, the powerful role of the Bureau of Reclamation in the West understandably led historians to focus on the storage and diversion of water for agriculture. The assumption that the history of water development in the West is a synonym for the history of irrigated agriculture in the West is persistent. In a recent overview of western history, making this assumption explicit, an item in the index reads, "Water supply. *See* irrigation."[4]

To the degree that urban water development has received the attention of historians, the focus has been on Los Angeles and San Francisco, respectively, on the diversions from the Owens and Hetch Hetchy Valleys. While California case studies make for illuminating comparisons, a study of Denver Water has the advantage of reacquainting everyone—Californians, Coloradans, westerners, and easterners—with the significance of the interior West and of ways in which California patterns both match and differ from patterns elsewhere in the region.

Studying Denver Water enhances our recognition of the importance of cities to regional history while also bringing to our attention a set of unexpected similarities between the eastern and western United States. To many observers of the West, the comparative scarcity of water has been the principal feature of regional distinctiveness. Every map of national precipitation patterns seems to make the case for the uniqueness of the West because of its much lower rates of snowfall and rainfall and much higher rate of solar-driven evapotranspiration. And yet a glance at the water systems of cities of the eastern United States delivers a sharp blow to this assumption of water-defined regional uniqueness. Very much like western systems, the infrastructures delivering water to cities like New York and Boston reach far into the rural hinterland, tapping into waters that are not by any definition riparian (that is, the cities draw on diversions from rivers that do not flow by the land occupied by the cities).

The island of Manhattan, for instance, was almost western in its scarcity of water, with a few streams and a number of springs at the time of European settlement. By the 1840s, the city of Manhattan undertook a precedent-setting transbasin diversion, drawing

The top photo shows the Hetch Hetchy Valley in the early twentieth century, before it was submerged, and the bottom photo shows the valley as it appears in the early twenty-first century from the top of O'Shaughnessy Dam. In both photos, Kolana Rock rises on the right, Tucculala Falls and Wapama Falls cascade down on the left, and the waters of the Tuolumne River occupy the center of the valley.

water from the Croton Reservoir, and this reach extended farther and farther into the state of New York over the next century. The remarks of New York mayor Philip Hone celebrating the delivery of Croton water to the city in 1842 had an anomalously western tone:

> Nothing is talked of or thought of in New York but Croton water; fountains, aqueducts, hydrants, and hose. . . . It is astonishing how popular the introduction of water is among all classes of our citizens . . . Water! Water! Is the universal note which is sounded through every part of the city, and infuses joy and exultation into the masses.

Boston traced a similar pattern with the construction of the Cochituate Reservoir in the 1830s. Whatever the local rates of precipitation might be, piling a bunch of people in a dense urban settlement will, in short order, create a need for more water than the immediate area offers. The urban acquisition of rural water resources and rural resentment of this intrusion form a pattern that the American West shares with the American East and, indeed, most of the planet.[5]

Along with a wider geographical range of reference, this book also invites readers to think in larger units of time, both before and beyond the present moment. The history of water development in Denver offers a particularly fine post for observing the astonishing and implausible workings of historical change and, in response, for cultivating an appropriate level of humility and modesty in our anticipations of our own unknowable future.

The success of an organization like Denver Water meant the frustration of the ambitions and hopes of other organizations, institutions, groups, and communities. This is a story in which success hinged on actions to claim water rights expeditiously, to overrule opposition from rivals, and to treat the interests of Denver residents as an unyielding priority. And yet this is also a story structured by episodic efforts at regional collaboration, at the use of negotiation instead of the sheer imposition of power, at the encouragement of thriftier uses of water, and at drawing on Denver Water's tradition of foresight to deal with vast and disorienting challenges like global climate change.

For all its efforts to practice foresight, the Denver Water Department itself has been tossed around by unexpected twists and turns of historical change. The rise of environmental movements and

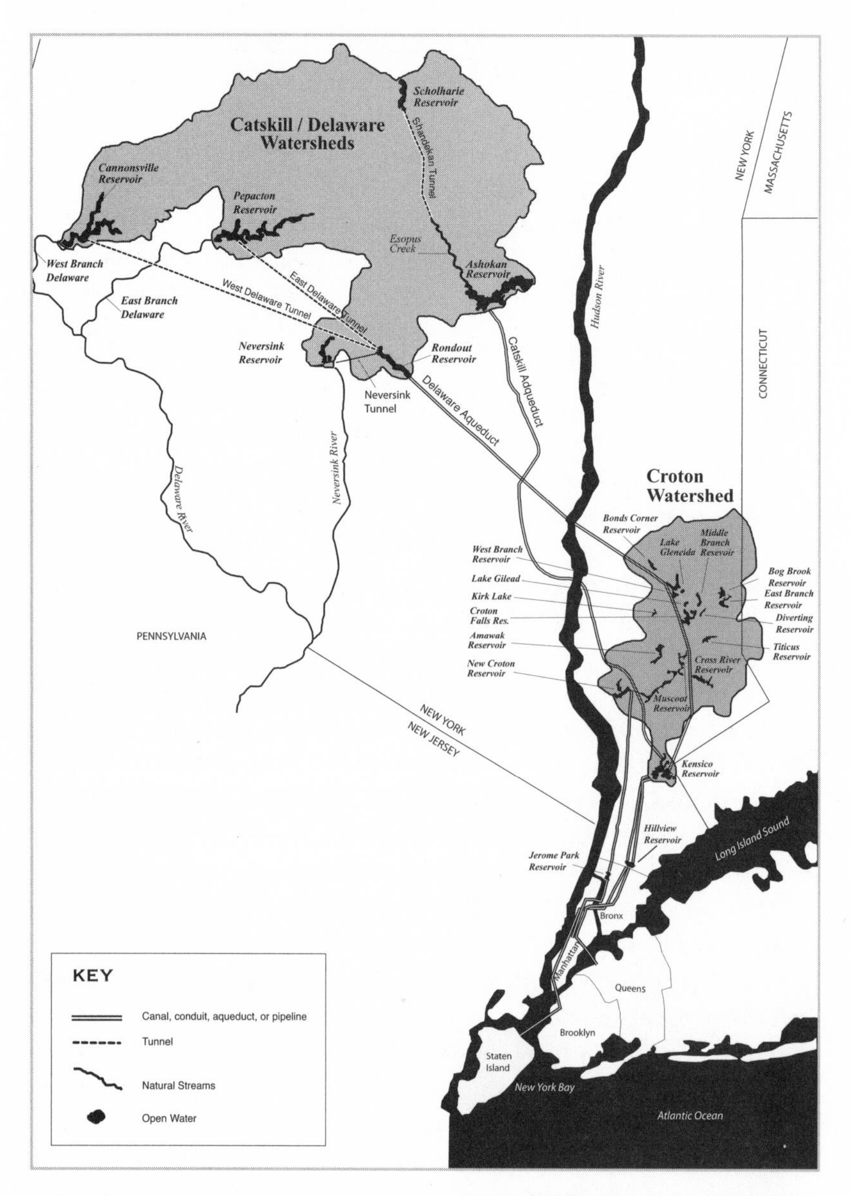

With a water supply system that reaches far out into its hinterlands, New York City has much in common with cities in more arid western locations like Denver.

Congress's passage of an array of forceful laws directed at environmental protection transformed the world in which the organization operated. In the environmental history of the twentieth century, a few celebrity events have registered as major landmarks in a changing configuration of power: the battle over a dam that was to be built in Echo Park in Dinosaur National Monument; the struggle over proposed dams in the Grand Canyon; the famed hit list by which the incoming administration of President Jimmy Carter called into question the future construction of federally sponsored dams. The Environmental Protection Agency's veto of Denver Water's proposed Two Forks Dam ranks with these other landmark events. Like them, the Two Forks veto is a crucial reminder that historical change is rarely a matter of linear progression or "more of the same," and far more a demonstration that contingency and choice interrupt and redirect seemingly well-established trends.

In the story of Denver Water, generations of people did their best to anticipate the future and to avoid the necessity of rushed and precarious action, aware that a ditch in time was preferred by far over last-ditch efforts. And yet the rise of environmental regulations and the related increase in the influence of Denver Water's opponents represent elements of historical change that the leaders of Denver Water did not see coming. Like many other agencies involved in natural-resource management, Denver Water scrupulously and wholeheartedly pursued goals that had been identified as progress. Abruptly, the definition of *progress* changed directions. In the manner of Dr. Jekyll turning into Mr. Hyde, projects that had once been perceived as advancing civilization, making the desert bloom, and supporting prosperity through growth were recategorized as intrusions and impositions that disrupted tranquil landscapes and natural harmony. Heroes of American development were recast as arrogant imperialists of human dominance.

While the fairness of these reappraisals has been and will continue to be a subject of vigorous debate, this shift in judgment nonetheless delivers one clear lesson of equal value to proponents and opponents of the massive campaign to rearrange the supply of water in the American West: in light of the improbability and unpredictability of change over time, the cultivation of nimbleness and the pursuit of humility serve as the necessary companions of ingenuity and foresight. Like the material infrastructure we inherit from the past, in light of the extraordinary and disorienting pace of change in our world today, ingenuity and foresight themselves require maintenance, reinvestment, and constant redesign. This book rests on the conviction that historical perspective can help in that cause.

Part 1 of this book examines the history of water in the Denver area, beginning with Indian occupation. Chapter 1 recounts the early efforts of American settlers in the Denver area to secure water for a growing city situated at the confluence of two streams, on the arid high plains, in the evening shadow of the Rocky Mountains. Chapter 2 chronicles Denver's expanding reach westward into the mountains for water and the campaign that led to the creation of the Denver Water Board in 1918.

Part 2 tells the tale of the Denver Water Board from its establishment to the present day, beginning with, in chapter 3, an essay on Glenn Saunders, a dynamic, colorful, and complex person who served as the public face and driving force of the Denver Water Board for much of the agency's middle age. Chapter 4 continues the story of Denver's quest for water from more distant mountain sources, diverting water from the headwaters of the Colorado River on the state's Western Slope under the Continental Divide and into the Denver Water Board's supply system. Chapter 5 takes on the complex events surrounding the Blue River and shows Denver Water at the height of its powers.

Chapter 6 tracks the rise of environmentalism in the second half of the twentieth century and its impact on the Denver Water Board, focusing on the contentious debate over the proposed Two Forks Dam and its eventual veto in 1990 by the Environmental Protection Agency. Chapter 7 assesses the degree to which the agency changed and adapted in the wake of the defeat over Two Forks, a period defined by the leadership of Chips Barry, manager of Denver Water from 1991 to 2010. The conclusion considers a number of common but mistaken assumptions about water, using the perspective provided by the history of Denver Water to arrive at better, more accurate understandings. And in the afterword, Chips Barry offers his own views and timely observations borne of two decades managing a water utility in the West.

There are reasons for a reader to start this book at the beginning and read straight to the end. But there are equally good reasons to skip around in the text. Writing about water history poses a strenuous literary challenge. The topic of water law, for instance, requires the writer to pay constant and close attention to providing sufficient clarity and simplicity without compromising accuracy. Readers encountering a tough patch—a description of a complicated engineering network of reservoirs, tunnels, and treatment plants, for instance—should exercise complete license to jump ahead to a livelier section. As a person lucky to have been born in an era before the diagnosis of attention deficit disorder was in play, I am the last person to complain if today's readers, cognitively reshaped by the digital and Internet age, find that hopping around a text is their preferred mode of taking in information.

As the book was nearing completion, I learned of the historians' enterprise known as envirotech—the merging of the fields of environmental history and technological history. A recent collection of articles by envirotech scholars explains this movement. Since "the relationship between people and nature is dynamic, interactive, complex, and messy," the result is that "nature and technology—and the way in which we understand the two—have become more and more entangled, blurring boundaries that once seemed so clear." As historian Martin Reuss sums this up, "Envirotech scholars attempt to unite human and natural history."[6]

As an inordinately sociable person, I am the equal of George F. Babbitt as a joiner; if I were to bowl, I would not imagine "bowling alone." In truth, if I encountered a club that would not have me as a member, I would probably still pick up a membership brochure. But I feel greater awkwardness at proclaiming myself a member of a scholarly movement that I learned of just as I finished a penultimate draft of this book. Still, having read some of the work in this emerging field, I can see an unmistakable match. My sense of kinship soars when I find a historian declaring, "Technology not only brings people closer to nature but also distances them from it. It can mask nature from human perception, as well as sharpen people's understanding of the world around them." When a pair of envirotech historians note that few people in industrialized countries "recognize their ties to the hydrologic cycle each time they turn on the tap," it is hard not to race forward to shake their hands and to declare, "The hope to change the situation you have described so well is what drives my book!" So, with the most honest acknowledgment of my latecomer status, I will go ahead and nominate this text for membership in the world of envirotech history.[7]

💧💧💧

By the late twentieth century, the writings of Edward Abbey had gained a status close to scripture for a widely held interpretation of water in the American West. "There is no lack of water here," Abbey said in a frequently quoted declaration, "unless you try to establish a city where no city should be."[8]

With a greater willingness to face up to the weight of history, water expert Peter Gleick provided a sharp rejoinder to Abbey's seemingly persuasive decree. Wiser practices in managing western water, Gleick has said, will not be a matter of "taking down Las Vegas or Los Angeles or Albuquerque or Phoenix and building them back somewhere where there's water. That horse has left the barn."[9]

To people who share Abbey's views, the Front Range of Colorado no doubt qualifies as one of those places "where no city should be." But what about those of us who, like Gleick, see little practicality in the project of wishing the West's cities out of existence? What are we to make of a city that plunged ahead and got itself established in a place with a characteristically Western "lack of water"?

Read on.

Part One

The Tangled Ties of Growth and Water

The West left settlers aghast;
It was dry; it was rugged; it was vast.
They thought water was the trigger
For making towns bigger,
An idea whose time is now past.

Chapter One

Engineered Eden

Here is a land where life is written in water,
The West is where water was and is.

—Thomas Hornsby Ferril, "Here Is a Land Where Life Is Written in Water"[1]

Those who favor our plan to alter the river
raise your hand. Thank you for your vote.
Last week, you'll recall, I spoke about how water
never complains.

—Richard Hugo, "Plans for Altering the River"[2]

Once upon a time, the area where Denver now sits was defined by water—in unmanageable and unimaginable abundance. The part of the planet we know as the Front Range of Colorado once sat "600 feet beneath the salty waves of a giant sea."[3]

Seventy million years can make quite a difference.

And so can a hundred years, when human ingenuity enters the picture. In the mid-nineteenth century, American settlers took on a place that seemed irreparably short of water and thus nearly uninhabitable and turned much of it green and heavily populated. Americans did this by acquiring and transporting water from distant streambeds and directing it to lawns, gardens, parks, sinks, bathtubs, swimming pools, fountains, farms, and factories. The places where the water arrived were obviously transformed, but so were the places from which it departed.

The perspective offered by geological change over the millennia reminds us that landscapes and ecosystems are undergoing constant change; the idea of the balance and stability of nature is more romance than reality. Meanwhile, the perspective offered by historical change over a century reminds us that human beings have extraordinary powers to accelerate the pace of change and that their successes and achievements face a big challenge in maintenance and duration. Both the flow of water and the flow of time can mock human enterprise and intention in very dynamic ways.

Historical perspective also reminds us that there has been enormous variation in the assumptions that human beings have made about how much water they *need.* Two hundred years ago, with bison herds providing crucial subsistence for Indian people, following the herds put a premium on knowledge of the area's rivers and springs. As they traveled through the plains and foothills, Cheyenne and Arapaho people needed to know where to find drinking water for themselves and their horses. While access to water was the key aspect of choosing a campsite, modest demands for water made it unnecessary to store or divert it from its original channels. Moreover, before American farmers, ranchers, and urbanites lowered the water table by pumping groundwater and capturing much of the surface runoff, pockets of surface water, in the form of prairie lakes and perennial springs, were common, so water could be found in shorter spans of travel. The banks of streams and rivers also supplied wood for the building of lodges and for fires for cooking and warmth, while river valleys, with their bluffs, banks, and canyons, provided refuge from the worst of winter storms.

Indian people thus had few reasons to think of their home as too dry and therefore unfit for human habitation. Still, as historian Elliott West put it, "Hunters had to fashion their living in the great spaces away from dependable rivers," and part of that adaptation was knowing the routes to find water quickly before hardship took hold. A satisfactory arrangement with water was not always in the picture for Native people. A period of drought and the resulting decline of grass could mean very tough times for bison and horses, and, therefore, for human beings. In the mid-nineteenth century, dry spells occurred several times, adding to the troubles of Indian people as they were confronting American invaders. Altogether, the contrasts between Indian customs and American customs are important reminders that variations in culture and economy produce very different definitions of how much water humans need and how much the world must be reworked to provide it.[4]

In 1806, on a journey of exploration that delivered more frustration than satisfaction, American explorer Zebulon Pike found unexpected promise in what he saw of western aridity. Traveling over the plains to the Colorado Front Range, Pike noted the absence of trees and the sparse flow of the streams and rivers.

What he saw did not hold hope for American settlement. And yet, despite our images of nineteenth-century Euro-Americans as instinctual expansionists, in a memorable passage that reminds us of the difficulty humans face when they strain to see the future, Pike described this limitation as good news:

> From these immense prairies may arise one great advantage to the United States, viz: The restriction on our population to some certain limits, and thereby a continuation of the Union. Our citizens being so prone to rambling and extending themselves on the frontier will, through necessity, be constrained to limit their extent on the West to the borders of the Missouri and Mississippi, while they leave the prairies incapable of cultivation to the wandering and uncivilized aborigines of the country.[5]

Fearing that a republic would be overstretched if it went past a certain size in territory and population, Pike saw in the limited water supply of the interior West one of the few forces strong enough to restrict a people who were certifiably "so prone to rambling and extending themselves on the frontier." Along with keeping the nation at a size that would be compatible with its system of governance, aridity also seemed to offer a solution to the dilemma of Indian-white conflict. If the plains proved unworkable for white settlement, then, by that very quality, they could provide a refuge for Indian people who would find no permanent place in better-watered parts of North America. As historian William Goetzmann observed, Pike's "view of the West set the popular pattern for many years to come." The interior West "was no place for the extension of civilization with farms and towns and mechanical pursuits. Rather the West . . . was a barrier which would contain the population and save the Union." This was not necessarily a misperception, Goetzmann explained further; in the early nineteenth century, the plains "were unfit for widespread settlement, given the level of American technology at the time."[6]

Thirteen years later, the Stephen Long expedition reached very similar conclusions. After a crossing of the plains in 1819, the chronicler of the expedition, Edwin James, concluded that the area would "prove of infinite importance to the United States inasmuch as it is calculated to serve as a barrier to prevent too great an extension of our population westward."[7]

An anachronism that would have wandered without a home in the nineteenth century, the term *growth control policy* did not appear in the vocabulary of these explorers. What Pike and Long saw in the limitations of the West's water supply was indeed something very much like a national growth control policy. In their judgment, the dryness of the West relieved Americans of the burdens of tough decisions. Using a constrained supply of water to carry its message, nature would act as legislator and decision maker and call a beneficial halt to westward expansion.

While not every observer joined Pike and Long in finding national benefit in the power of aridity to obstruct expansion, many matched their impression of the sparse natural assets of the plains landscape. In 1846, Boston Brahmin Francis Parkman traveled through the West, accumulating experiences that would shape his later prominent career as a historian. Just as much as trappers extracted beaver pelts and miners would extract minerals, East Coast writers like Parkman extracted literary nuggets from their travels and delivered them to markets back home. The West's mountains gave Parkman all the right opportunities to celebrate the romantic appeal of the landscape, but the plains made a writer's job harder. Taking in the country he saw as he traveled from Fort Laramie to Bent's Fort, Parkman was pushed past the borders of that excellent advice: "If you can't say something nice, then don't say anything at all." The best that Parkman could do with the discouraging landscape was to describe it, in his influential book of 1849, *The Oregon Trail,* as an "arid desert" where "the only vegetation was a few tufts of short grass, dried and shriveled by the heat."[8]

Far more preoccupied with a search for water than a search for profit-delivering resources, people in Parkman's situation urgently inspected the landscape for signs of moisture. The West had water, he and many of his contemporaries knew, but it was often the wrong kind of water (mineral laden and undrinkable) or sometimes the right kind of water but located in the wrong place, not at hand when people most needed it. It was a happy, and rare, circumstance when the water presented itself in a manner that matched and suited the human desire for it. On one occasion in his journey, Parkman and his thirsty companions thought they were approaching a large stream. To their mortification, they found it to be completely dry. Parkman described the dismal scene: "The old cotton-wood trees that grew along the bank, lamentably abused by lightning and tempest, were withering with the drought, and on the dead limbs, at the summit of the tallest, half a dozen crows were harshly cawing, like birds of evil omen." The group went on to find enough water in the South Platte to make it hard to ford, and then they came upon Cherry Creek near the spot that would later come to be called Denver. "The stream, however, like most of the others which we passed," Parkman observed, "was dried up with the heat, and we had to dig holes in the sand to find water for ourselves and for our horses." Soon after, excess replaced scarcity, and the group was drenched by a tumultuous rainstorm. In the proportions of water present at the eastern base of the Rockies, too much could replace too little in the course of a few hours.[9]

Americans arriving in the interior West from eastern points of origin traveled heavily laden with cultural baggage. A common element in that baggage was a preference for landscapes shaped by the humid climates they had left behind. By this preference, the western landscape, without lush vegetation, was disappointing and deficient. Thus, for four decades, as maps labeled the plains as the Great American Desert and overland travelers echoed the explorers' opinions of the unappealing strangeness of the arid lands, Pike and Long seemed to be pretty capable practitioners of foresight. But then a factor that had never entered into their minds lowered their performance as prophets.

The discovery of gold in the Rockies unleashed a rush into the region. Even though American settlement remained precarious in the first years after the rush, Pike and Long's prediction was soon looking shaky and unconvincing. In hindsight, it is clear to us that seasonal constraints on exploration and travel had led to a misappraisal of the water resources of the West. Waiting for the growth of grass for their livestock before leaving the Mississippi and Missouri Valleys, nineteenth-century travelers rarely reached the foothills of the Rockies in time to see the heavy spring runoff, produced by the melting of the snowpack, which in many years filled streambeds and sometimes overflowed banks. By the time the explorers and travelers arrived on the scene in the summer, the runoff was far downstream, headed to the oceans, and many of the so-called rivers and streams of the plains, even close to the foothills, seemed shrunken and paltry. The enormous expanses of plains between the rivers, combined with their diminished state in summer, had fostered the impression that the region was irreparably short on water.

The founding and growth of the city of Denver seemed to invalidate these earlier predictions of intrinsic limits to the settlement of the region. Pike and Long, after all, had understood *American settlement* to mean "farms," and a future role for cities had not been on their minds. And, just as they had not anticipated a mineral rush that would lead to a concentration of settlers in towns, they also had underestimated the extraordinary force of human ingenuity and enterprise.

On July 7, 1858, a prospecting expedition led by William Green Russell of Georgia, working along the Front Range of the Rocky Mountains, discovered placer gold near the mouth of Dry Creek as it empties into the South Platte River in present-day Englewood.[10]

This was not the first discovery of gold in this area, at the time part of Kansas Territory, but it proved to be the best publicized and therefore the most consequential. On August 26, 1858, the *Kansas City Journal of Commerce* announced the news: "THE NEW ELDORADO!!! GOLD IN KANSAS!!"[11]

The word spread, and in the spring of 1859, close to one hundred thousand people headed for the part of Kansas that would be organized as Colorado Territory two years later.

In the American West and many other parts of the world, mineral rushes dramatically accelerated the pace of migration into areas that had previously been judged unsuitable for settlement. The discovery and extraction of minerals led to the immediate founding of towns. As merchants realized the opportunities presented by mining the miners, what historian Gunther Barth called "instant cities" came into being by trade and commerce.[12]

Where the plains met the Rockies, new arrivals seized the opportunities that might converge for a town placed at the base of the mountains. The creation of the town was far from a thought-out and organized project. Entrepreneurs founded the towns of Auraria and Denver on opposite banks of Cherry Creek in 1858. These two communities competed with each for a spell, until Denver gained the upper hand and engineered a merger in April 1860. In a fine example of the long-lasting consequences of decisions made in the early days of a community, the streets and lots of the towns had been laid out with grids set at different angles, and the split origins of the city of Denver have sent numerous visitors off on indirect and unintended journeys as they tried to navigate their way through those two misaligned grids.

Like many residents of western towns, early Denverites found themselves in precarious and tenuous living conditions. In April 1863, fire destroyed most of the business district. This was a very common pattern of western towns, to the point that extraterrestrial observers would have puzzled over the motivation and strategy of the peculiar species that bustled around piling up wood in a central location, igniting those piles, and then experiencing apparent regret and distress when the wood burned. Efforts to avoid such misadventures led to both city regulations requiring the replacement of wood with brick in business districts and to the recognition that, when fire threatened, a reliable water supply could prove to be an essential element of the safety of the community.

If Denverites had unintentionally but very effectively put themselves in the line of fire, they had intentionally placed themselves equally in the line of water. Settlers lined the banks of the South Platte and Cherry Creek, near the point where the two rivers converge, with homes, churches, and businesses. To claim and hold a central place in the community, the *Rocky Mountain News* went so far as to build its offices in the sandy bottom of Cherry Creek's dry bed. In May 1864, heavy rains caused the creek to rise. With astonishing speed, the deluge destroyed the *News* offices and sent machinery, including a three-thousand-pound iron press, downstream. Five newspapermen sleeping in the building's second story awoke to a terrific "roaring noise" and saw the flash flood charging down the creek bed. As the building washed away, onlookers threw the trapped men some rope. The five barely escaped with their lives. Others were not so fortunate. Twelve people drowned as Cherry Creek and the South Platte River submerged neighborhoods, toppled buildings, and did its best to wash away the fledgling settlement.[13]

Soon after the flood, an observer found a warning in and drew a lesson in restraint from the flood. "Men are mere ciphers in creation," reflected Professor O. J. Goldrick in the *Daily Commonwealth and Republican.* "Had we continued thickly settling Cherry Creek as we commenced, and thoughtless of our future, see what terrible destruction would have been our doom, in a few years more, when the water of heaven, obeying the fixed laws, would rush down upon us, and slay thousands instead of tens!" Rhetorically awash, Goldrick was nonetheless engaged in the important project of drawing

In 1860, the *Rocky Mountain News* briefly let honesty compromise its usual unrelenting boosterism by describing Denver's landscape as "treeless, grassless, bushless." As this photograph of the paper's office in the early 1860s shows, the writer was not exaggerating.

In May 1864, a flash flood roared down Cherry Creek and inundated Denver, killing twelve people and washing away numerous buildings, including the *Rocky Mountain News* office.

a practical lesson from calamity. To at least a limited degree, newcomers were trying to understand and deal with their changeable relationship to water. After the 1864 flood, the Denver City Council asked the legislature for power to regulate building in Cherry Creek. The legislature granted the power the following year, and the city outlawed the construction of houses or businesses in the creek bed.[14]

Despite the occasional flood, scarcity of water posed a more-persistent challenge. Visitors often noted the town's lack of trees and other greenery. Stopping in Denver in the summer of 1859, journalist Albert Richardson dismissed the aspiring city as "a most forlorn and desolate-looking metropolis" situated upon a bleak setting of "low sandy hills entirely destitute of trees and with thin ashen grass dreary enough to eyes familiar with the rich green prairies of Kansas and Missouri."[15]

Even the *Rocky Mountain News* briefly let honesty compromise its usual unrelenting boosterism by describing the landscape in 1860 as "treeless, grassless, bushless."[16]

And yet, to Colorado's early leaders, the landscape was charged with promise. Colorado's first territorial governor, William Gilpin, led the pack of optimists. Where others saw only arid deserts, Gilpin saw a pastoral paradise waiting to happen. In his writing, Gilpin's verb tenses broke the speed limit, casting a possible future as an actual present: "These PLAINS are not deserts, but the opposite. . . . They form the PASTORAL GARDEN of the world." Few have matched the confidence of nineteenth-century boosters, and Gilpin thus saw, taking shape in the "treeless, grassless, bushless" terrain around him, "the auspicious cosmopolitan site of DENVER." Unlike the early explorers, Gilpin noted the seasonal shifts in the rivers and streams: "During the melting of the snows in the immense mountain masses on the western frontier of the Great Plains, the rivers swell like the Nile." He did not, however, propose the damming and storage of the spring runoff; on the contrary, frequently mystical in his interpretations, he saw the heavier flows of the spring as "yield[ing] a copious evaporation," which in turn produced "storm clouds" and thus "vernal showers." And, like many who would follow him, Gilpin found much to celebrate in the charms of a semiarid climate: "The atmosphere of the *Great Plains* is perpetually brilliant with sunshine, tonic, healthy and inspiring to the temper."[17]

As "a focal point of impregnable power in the topographical configuration of the continent," Gilpin's Denver was going to become the capital of commerce of the United States and, given its fortunate location as the midpoint between Europe and Asia, of the entire world. Doubt rolled off Gilpin's mind like water off the rocks above tree line:

> The scientific writers of our country adhere with unanimity to the dogmatic location somewhere of "a great North American desert.". . . Yet there is none, either in North or South America; nor is the existence of one possible. On the contrary, the least fertile portion of our continent is the siliceous slope of the Atlantic States, whose climate is also the most inhospitable.[18]

Having drawn a satisfactorily invidious comparison pointing out the crummy soil and discouraging climate of the East Coast, the territorial governor and booster of the first order returned to making his case for the glorious future of Denver and Colorado.

In hindsight, Gilpin's optimism might strike readers as comical in its wild ambition. And yet hindsight mockery runs aground on the fact that Gilpin, in a number of ways, turned out to be right. Thanks to astonishing exercises of human ingenuity, Denver—and many other communities in the arid West—turned into green, tree-filled places that millions now call home. It would be easier to enjoy a good laugh at Gilpin's expense if westerners did not see the realization of his vision wherever we look.

Just as William Gilpin was whipping up a global destiny for Denver in which water supply would pose no hindrance, promoters of expansion on the plains were arguing that settlement would in itself produce a happy change in the climate. Rain, they predicted, would follow the plow. The act of breaking and tilling the soil of the arid plains would coax rain to leave the heavens for the earth. The rain would then evaporate and return to the ground in later showers, in a pleasant cycle operated and maintained by the laws of nature. One of the theory's greatest supporters, Charles Dana Wilber, mobilized his training as a geologist to promote settlement in railroad land holdings. Wilbur saw no limit on the power of human will:

> Rain follows the plow. . . [I]n reality, there is no desert anywhere except by man's permission or neglect. . . . The Creator never

Colorado's first territorial governor and booster-in-chief William Gilpin looked out at Denver's dusty setting and envisioned a pastoral paradise waiting to happen. His ambitious prophecies might sound laughable today if they had not been so fully realized.

> imposed a perpetual desert upon the earth, but, on the contrary, has so endowed it that man, by the plow, can transform it, in any country, into farm areas . . . the power [is] in our hands to make the wilderness and waste places glad, and to make even a desert blossom as a garden with roses.[19]

If rain would be cooperative and follow the plow, then the demands on muscle and ingenuity would shrink wonderfully, making dams, reservoirs, ditches, canals, and headgates superfluous and unnecessary. And, for a while, a series of wet years seemed to validate this cheerful theory.

Denverites did not, however, sit waiting for precipitation to deliver the water they wanted to the sites where prospective users awaited it. If the city was going to grow to meet the founders' ambitions, they knew they needed a system to transport water from the sites where nature had apparently misplaced it.

♦♦♦

The riddle posed by tracking the headwaters of a river and locating its point of origin tested the wits of any number of explorers of the American West. Following the instructions given to them by Thomas Jefferson, Meriwether Lewis and William Clark took on this conundrum with the Missouri River. As they slowly made their way upstream, the proliferation of converging branches of the river made it difficult, even impossible, to identify the main stem that they were supposed to ascend. Water from springs, rainfall, and snowmelt comes together in trickles; trickles carve out channels and merge with other small tributaries; and finally a sufficient number of these sites of confluence add up to an unmistakable river. Rather than one clearly marked point of origin, a river literally descends from many sources.

Tracking the convergence of the enterprises that finally came together under the name Denver Water presents a similar challenge to the explorer of history. A bewildering array of competitors emerged to supply the growing town with water. In just a few decades, the city and its water companies grew exponentially, rushing into the future with a momentum that could match the flow of fast-moving Rocky Mountain streams in spring.

Between 1859 and 1918, at least a dozen water companies came into being in Denver and its immediate surrounding area. They competed and sometimes consolidated with each other, following the preferences of businessmen, politicians, and the local citizenry. The complexity of the story requires a direct and reassuring address to the reader: do not panic. You are about to encounter a flurry of names. Do not burden yourself by trying to commit to memory each one of these; there will be no test. But come to attention when you see the words Citizens' Water Company and Denver Union Water Company.

In 1859, just months after the founding of Auraria and Denver, local citizens incorporated the first water company. Calling itself the Auraria and Cherry Creek Water Company, this organization raised hopes that a serviceable ditch would soon be carrying water to the town. But the plan, and the company, delivered no results. In the meantime, following the almost universal pattern of new American settlements, residents took advantage of underground water and relied on private wells.[20]

In 1860, a second company came into being. The Kansas Legislature (Colorado would not gain its own territorial status until 1861) granted the Capitol Hydraulic Company a "perpetual charter" to build a ditch from the South Platte, conducting (the theory went) clean and abundant water to the east side of Denver. The Capitol Hydraulic Company began construction in the summer of 1860. Though tiny in scale compared to many later water projects, the ditch represented an important milestone in the transformation of Denver.[21]

The company was, to put it mildly, well-connected, indicating the importance of water in the plans of Denver's leaders. A. C. Hunt, who later became territorial governor, served as the company's first president. William Byers, the very influential publisher and editor of the *Rocky Mountain News*, used his publishing platform to promote the project aggressively, asserting that "no work has ever before been undertaken of such vast importance to Denver. . . . Gardens and farms can be irrigated along its course in all the space between it and the river, and an unfailing source of water for all emergencies will be furnished to the whole city." As an artifact steeped in confidence and cheer, Byers's declaration set the terms for many such declarations in the future development of Denver and the West; an "unfailing source of water" was an ambitious phrase to use in a region already so well known for its uneven precipitation and stream flow. It was also illuminating and illustrative to see the way in which Byers had blended urban uses and agricultural uses; "gardens and farms" were presented as congenial elements in "the whole city."[22]

Applauded, celebrated, and much anticipated, the Capitol Hydraulic Company's construction of the Big Ditch (as Denverites took to calling it) was, in practice, more plodding than triumphal. Faulty calculations of the canal's gradient delayed the project in early 1861. Later that summer, the economic disruptions and political tensions of the Civil War put brakes on the already slow process. Between national disunion and the exhaustion of the supply of easy placer gold, immigration into the territory ground to a halt and even pulled some back East. Discouragement and trouble came in multiple forms: the fire that destroyed the city's business district in 1863; a massive infestation of grasshoppers that same year; and the devastating Cherry Creek flood of 1864.[23]

Concerns about fire and drought prompted the people who stayed to push for a more reliable water supply to fight fires and irrigate lands, and the *Rocky Mountain News*, now out of its precarious creek-bed locale, took up the cause. John W. Smith, a wealthy Denver businessman and entrepreneur, responded in 1865 by buying out the Capitol Hydraulic Company and reviving the construction of the Big Ditch under the auspices of his Platte Water Company. Smith moved the ditch's headgate upstream to secure the necessary slope for water to flow by gravity and used steam plows to speed up the work. Finally completed in 1867, the Big Ditch stretched twenty-four miles from Platte Canyon to Brown's Bluff (now known as Capitol Hill) in Denver, creating Smith Lake as a repository out of an old buffalo wallow (where both old and young buffalo once wallowed) in what is today Washington Park. Diversion ditches fed from the main channel, irrigating farms on the town's outskirts and running through small, uncovered canals alongside city streets in a distribution system that permitted city residents to water their yards and gardens.[24]

Reasoning that a resource so crucial to the public good required the exercise of public authority, the editors of the *Rocky Mountain News* and *Denver Tribune* urged city government to purchase the water system for fighting fires, beautifying the city, and supplying domestic water. In 1866, the territorial legislature formally authorized Denver "to provide for and regulate the manner of introducing water into the city for irrigating and other purposes." Coordinating those two roles—"providing for" water and "regulating" its introduction—would, over the next century and a half, provide strenuous exercise for the officials thus obligated to get the residents the water they believed they needed, while also trying to regulate the urgency and scale of that need.[25]

The city granted its first irrigation franchise to John Smith's Platte Water Company in April 1869.*

By the following spring, the company was providing water throughout Denver. Initially, the city paid the Platte Water Company $7,000 per year for "a daily supply of 400 square inches of water for six months." But after a series of disagreements between company officers and city administrators, Denver's leaders concluded that it made more sense to purchase the ditch outright. Empowered by the territorial government in 1874 to own its own municipal water system, Denver negotiated to buy the Big Ditch from the Platte Water Company for $60,000. In 1882, the city made the last payment on the buyout agreement, and Smith's ditch was rechristened City Ditch (although a generation of Denverites stuck with habit and continued to call it the Big Ditch).[26]

The completion of the Big Ditch had an instant and lasting impact on Denver's appearance. The *Rocky Mountain News* called the ditch "the turning point in our city's history." Suddenly the visions of the boosters seemed possible, even probable. The secretary of the Denver Board of Trade, Henry Leach, was a man overcome with cheer: "Denver will rival the great manufacturing cities of the East; conduct a portion of [the water] through the streets and into our houses, and Denver will become the most beautiful city on the continent." In short order, such predictions began to gain substance as the city's suntanned landscapes gave way during the late 1860s and the early 1870s to green lawns and ornamental flora that remained lush all summer long. By 1889, a visitor, Rezin Constant, could report that "there are the most exquisitely beautiful lawns in Denver I ever saw; great varieties of Roses . . . in full

* As Earl Mosley describes the franchise, "It granted to the Platte River Company the privilege and right of way in all the streets in the City of Denver, to build ditches, flumes, and viaducts for the purposes of conveying water from the said company's main ditch, through the City of Denver for irrigating and other purposes. . . . All persons were given the right to contract with the company for the use of water, and a sliding scale of fines was set up for use against any person for turning water out of ditches, without first having obtained permission to do so from the Company." Mosley, *History of the Denver Water System*, 98.

bloom; Fuschias, in short, every desirable flower." Back in 1863, Presbyterian minister William Crawford, encapsulating the dismay Americans felt when they contemplated Denver's setting, "was overcome" at a church service "by the urge to have the congregation sing a hymn entitled, 'Lord what a wretched land is this!'" Just a few years later, the tune had changed dramatically.[27]

Agricultural irrigation systems began to transform the land around the fledgling city from dry prairie into productive farmland, providing fresh food for residents of the Front Range and resolutely challenging the image of the region as a desert. When Albert Richardson, who in 1859 had described Denver's setting as desolate and dreary, made a return visit in 1865, he found that "Colorado agriculture was already successful," supplying "the population of the territory with every farm product except corn." A reporter for the *New York Tribune* was similarly impressed by the area's transformation: "I verily think that if those who six years ago saw nothing but arid hills and fields of cactus, forbidding cultivation, could behold some parts of Colorado at present, they would open their eyes in astonishment. . . . I am fast inclining toward the opinion that there is no American Desert on this side of the Rocky Mountains." The writer William Makepeace Thayer reveled in before-and-after contrasts, comparing his visit in 1859 to his return in the mid-1880s. "The most marvelous growth of modern times," he declared, "is the city of Denver, Colorado." The grim landscape where pioneers had once lived in humble tents and cabins now held "the largest, richest, and most beautiful city of its age on earth—a sparkling, costly jewel on the bosom of the 'desert.'" Thayer continued:

> The city is handsomely laid out, with wide avenues lined with shade-trees and beautified with irrigating rivulets; large and costly warehouses and public buildings, street-cars; the electric light; water-works; elegant churches; newspapers, and schools unsurpassed by those of Boston; telegraph, telephone, and railway facilities; in short, everything necessary to promote the growth of a marvelous city, which may contain, in twenty years, a population of two hundred thousand.

Thayer's readers could easily understand that the enhanced supply of water had made possible every dimension of this urban achievement, especially in the opportunities presented for an expanding population.[28]

As quickly as they reconfigured their landscapes, the residents of Denver were equally quick to take their new circumstances for granted. The availability of more water brought into being, almost instantly, many new "needs" for it. It suddenly became possible to plant and maintain a host of nonnative grasses and plants for lawns and gardens. "Sprinkling wagons" materialized on the streets, offering to dampen "the dusty roads in front of [business and residential] properties at 25 cents a week for a 250-foot front." As local historian Louisa Ward Arps wrote, with ditches to drink from, "thirsty dogs and horses were delighted. Trees grew, gardens flourished, and lawns were praised for their greenness all summer long."[29]

A wildly transformed landscape soon began to look normal, and a desire for a conventionally pretty landscape quickly got reclassified as a need.

♦♦♦

"Water for irrigation is abundant and good, but it is not fit for household use," the *Rocky Mountain News* complained in 1870. Even after the completion of the Big Ditch, many of Denver's residents still hauled their drinking water home in buckets or barrels directly from the Platte River or obtained it from private artesian wells. And, according to William Byers at the *News*, they were hoping for better arrangements: "The people are getting interested and importunate, because all are satisfied that an ample supply of water must be secured." Not enough water was flowing through the Big Ditch and into the street-side canal network. Quality and quantity were proving to be connected: the quality worsened when the quantity diminished.[30]

Satisfaction with the water supply rarely achieved much in the way of duration. Two years after the city had granted a franchise to the Platte Water Company to provide water to every street in Denver, newspaper editorials had returned to expressions of concern and discontent about the city's water supply. This time, the concern shifted toward problems of quality, with one newspaper declaring that a system capable of providing reliable water quality, not just quantity, was the "most important improvement now required for Denver." In a June 1871 editorial carrying the archetypal western declaration "We Want More Water," the *Rocky Mountain News* complained that "the ditch company are [*sic*] not furnishing what water they agreed to and confidently assured the Council they would be able to supply. If they knew their ditch was unequal to the job they should have said so."[31]

Byers felt that the Platte Water Company should be held to its promises for water delivery, regardless of variations in precipitation. While the company promised in 1870 to provide 400 inches of water per day, the actual daily average in June 1871 was 275 inches. "Possibly the Platte is low," the 1871 *News* editorial continued. "Indeed, it is notable that it is so; but this ought not to prevent the ditch company from perfecting some arrangements for forcing some water through their ditch and into the city." Denverites had become tough and demanding customers, exhorting the city's major water supplier to "do something . . . and give us more water during the hot weather."[32]

A local entrepreneur saw opportunity in these concerns. Railroad promoter and businessman James Archer had already founded the city's first gas company in January 1870 when, ten months later, he started the Denver City Water Company. While the Platte Water Company delivered water that traveled through a series of open street-side canals and was only of sufficient quality for irrigation, the Denver City Water Company became the first water supplier in Denver to deliver water via pressurized underground plumbing for domestic use and firefighting.

As president of the Denver City Water Company, Archer recruited other important citizens into his organization. These strategically selected backers held positions at banking institutions: David Moffat, Walter Cheesman, Jerome Chaffee, and Edward M. McCook. The financial and political connections of these men secured a respectable initial capitalization of $250,000 for the ambitious Denver City Water Company. At the time, McCook was particularly prominent and well-connected, serving as territorial governor. Typically for their times, the careers of these men did not permit a clear drawing of lines of specialization. They took up a range of industries and enterprises—banking, railroads, real estate, municipal utilities—sometimes sequentially and sometimes simultaneously.[33]

The city council granted the company express permission to build a pumping station and to take water from the South Platte. With construction of the water works beginning in 1871, the Denver

City Water Company installed a pair of the admired Holly pumps in a station, linked to "a system of pipes laid to supply the central part of town." Digging a well into underground gravel beds in the South Platte, just below the mouth of Cherry Creek, Denver City Water Company crews connected the phases of Denver's rough-and-tumble water history by finding remnants of the *Rocky Mountain News* presses that had been swept away from the paper's creek-bed office during the flood of 1864. By 1872, the Holly pumps distributed water from this newly excavated well to taps all around the city.[34]

The pressurized system of underground pipes represented a significant advance over the paltry and often dirty summer flows delivered by open, street-side canals. The Denver City Water Company quickly surpassed the Platte Water Company as the city's major water supplier. As well as providing purer drinking water, called Holly water by Denverites, directly to household taps, the system also supplied water for fire hydrants. Between the hydrants and the city's Brick Ordinance outlawing wooden buildings downtown, no fire ever again equaled the devastation of the 1863 business-district fire.

💧💧💧

Like the Platte Water Company before it, the Denver City Water Company did not get to bask in the accolades of a grateful citizenry for long. Concern over sanitation soon refocused attention on water quality. Less than two decades after Denver came into being near the South Platte River, the city's principal source of water had become contaminated as the growing population overburdened the river's ability to carry away unpleasant substances. By the later half of the 1870s, Denverites frequently complained about the "foul appearance, taste, and odor" of their drinking water. With no sewer system, human waste from privies seeped into the surrounding groundwater and eventually into the South Platte River. Household and commercial trash, including discarded meats and vegetables, floated in the city's uncovered ditches into the river, just upstream from the Denver City Water Company's pumping plant. With that plant located in the middle of the city, the water pumped from the river had abundant opportunities to pick up sewage and waste from drainage ditches, as well as seepage from privies into shallow groundwater.[35]

Attempting to explain "the occasional appearance of dirty water in our pipes" in 1879, the Denver City Water Company blamed unusually heavy water flow that "stirred up a sediment that had occurred months past, and which so soon as it appears in the service pipes is washed out by flushing our mains." Nothing was wrong, the company's officials maintained. They "did not consider it worthwhile to reply to the various newspaper articles that have appeared within the past two weeks on the 'sickening impurities of the Holly water.'" They did, however, eventually acknowledge the mounting complaints by dispatching Professor Richard Pearce, a chemist, to test the water. Pearce backed his employers' position, stating that although the water "showed a slight turbidity, due to the presence of fine, flaculent vegetable matter in suspension. . . , [it is] among the first quality waters for drinking purposes." Despite Pearce's conclusion (itself filtered through the headline writers of the *Rocky Mountain News*) that "We Have the Best Drinking Water in the World," the report did not quiet public concern. Numerous citizens fell ill with typhoid, a fearsome illness spread through water contaminated by sewage, while Pearce worked away at his unconvincing report. Data disputes between experts had made an early entry into the Denver water scene.[36]

In a pattern shared with many other American cities, Denver residents suffered through at least six major typhoid epidemics between 1879 and 1896. During the 1879 outbreak, during which the fever afflicted more than six hundred people and killed at least forty, one local doctor surreptitiously gained access to one of the Denver City Water Company's reservoirs. The scathing description he wrote for the *Rocky Mountain News* about the facility's unsanitary conditions further undermined the company's protestations of purity:

> To this little reservoir . . . the necessarily foul water . . . flows into a small ditch from the mill run stream. In the little ditch I saw broken slop buckets, etc., detained by the shallowness of the water there. That water was perceptibly not wholesome as drinking water, yet it flowed under the edge of the [sluice] box and [into the river].

In the absence of antibiotics, the responsibility for the prevention of typhoid outbreaks fell largely upon the private companies providing water. Not all of these companies were up to the challenge, and the impact of sensational reports about unhealthy conditions of the water-supply system galvanized demands from citizens, physicians, and local newspapers for a proper sewer system and for governmental regulation to see that water was pumped from sources free from pollution.[37]

As early as 1867, Denver's water suppliers had already begun to change their practices to address issues of water quality. The earliest methods involved little more than trying to isolate the city's water supplies from the effluence of privies and farm animals. The companies built protective wooden coverings over reservoirs, and the city banned hogs within city limits. When the typhoid outbreaks exposed the ineffectiveness of these simple methods and citizen anger threatened revenues, companies began exploring the emerging technology of filtration as a more scientific remedy for contamination. In 1884, the first of many different filtration techniques went into practice in Denver; by 1889, the Platte Canyon Filtration Plant (later renamed the Kassler Water Treatment Plant) was removing impurities from water by sluicing it through beds of sand. Sand filters required a large tract of land, as well as a substantial labor force to clean the filters continually. But the effort paid off in lower rates of disease. The benefits of filtering were complemented by the advent of a typhoid vaccine in 1897 and by the widespread chlorination of water supplies roughly a decade later. Typhoid outbreaks grew blessedly rare in Denver and in most other American cities during the early twentieth century.[38]

♦♦♦

The sorrows of disease also galvanized the city's water providers into seeking better sources of supply. Even as Denver City Water Company president James Archer defended the purity of his company's water against the attacks printed in the *Rocky Mountain News*, he was already working to secure a cleaner and more abundant water supply. Archer and his company pioneered a strategy that later generations of city water leaders would follow: seeking a new water supply, they would extend their reach to sources that lay farther away from the pollution of populated areas, and they would capture those sources with bigger, more ambitious structures.[39]

In 1878, the Denver City Water Company formed a subsidiary, the Denver City Irrigation and Water Company, for the purpose of identifying a "higher source of supply and a more distant and less central location" for new, expanded waterworks. The company quickly constructed a reservoir (really a glorified canal big enough to store water) linked to the South Platte, which they named Lake Archer as a memorial to the now-deceased James Archer. (As Louisa Ward Arps noted, such gestures of commemoration can be uneven in their results: "It was named Lake Archer so that Denver would always remember its water pioneer. But the lake has dried up and who remembers Colonel James Archer?") Four Holly pumps, with a combined output of 5 million gallons of water a day, siphoned water from the new lake and pumped it to people's homes. The new West Denver Station (located near the Denver Water Department's present-day headquarters on 12th Avenue and Shoshone Street) was in operation by May of 1880. The company now held the capacity to deliver 9.5 million gallons of water per day, just as the city completed its modern sewer system and thus resolved some of the problems of water quality that had been plaguing residents. By 1881, the company had added two more pumps to the facility, increasing the station's pumping ability to more than 14.5 million gallons of water a day from the South Platte.[40]

"No anticipation then existed here of the astounding growth of the city that set in with a bound very soon" after these improvements, historian Jerome Smiley noted in 1901, "and the location of the new pumping station was thought to be far beyond the possibility of the city's encroachment." As Smiley's comments indicate, the city's expansion would render the best plans of the 1880s obsolete by the early twentieth century. With an enhanced water supply, the city grew, quickly closing in on the pumping station. Lake Archer, meanwhile, filled in with "silt and decaying vegetable matter" in just a few years. Water quality began another slump.[41]

With the Denver City Water Company now merged with its subsidiary, the newly named Denver Water Company set out to seek another water source even farther up the South Platte River. The company constructed wooden galleries below the riverbed of the South Platte to collect naturally filtered water as it seeped through layers of gravel and sand. Forty-eight-inch wooden conduits transported the clean water from the galleries to a new reservoir near the "West Denver" pumping station, which then, in turn, distributed it to at least marginally more healthful residents.[42]

♦♦♦

Since the troubled times during the Civil War, Denver's leaders had aspired to and reached a series of goals. The city had linked itself to the transcontinental railroad in 1870, and after statehood in 1876, Denver officially secured its status as the state capital in 1881. The city's growth made its boosters look more like solid prophets than men whose judgment had been weakened by unreasonable optimism. The population swelled from around 4,800 in 1870 to more than 35,000 in 1880, and continued to increase to nearly 107,000 in 1890. As the Denver Water Company struggled to keep pace with this growth, a number of ambitious entrepreneurs saw opportunity in the company's difficulties.[43]

Competition among companies took place in an uncertain and changeable relationship with city government. The experience of the somewhat ephemeral Denver Aqueduct Company offered one example. Founded in 1871, the Denver Aqueduct Company requested a right-of-way from the city in order to build a second ditch (the Big Ditch was the first) to the city. The *Rocky Mountain News* enthusiastically endorsed the proposal: "The construction and completion of this ditch will ensure one thing greatly needed, and what we have often urged, to wit: competition. With two ditches, our supply of water can hardly ever fail." The *News* may have been excited by the prospect of competition among water suppliers, but the city council's Water Committee was not similarly swept away. The committee postponed any decision on the right-of-way request for six months. When the time for decision came, the city council decided that, since precipitation was above average that year, the securing of a larger water supply could be put off again.[44]

This stuttering pattern of unreliable encouragement stymied the efforts of local entrepreneurs to challenge the dominance of the Platte Water Company and the Denver City Water Company. The Denver Aqueduct Company was not the only victim of the city council's changeable moods. The council initially provided financial support for the Denver Artesian Company's efforts to drill a municipal well, and drilling began in 1870. But drilling stopped in 1871 with an interruption of funding, and it was two more years before the promised money materialized and drilling resumed. When the company asked for more money, the city council balked, and the project finally failed. In the 1870s, the proposition that

From 1870 to 1890, Denver's population mushroomed from 4,800 to 107,000. Water providers struggled to keep pace with the rapidly growing city, shown here in the early 1880s.

competition among water companies would work to the public's benefit received much lip service but produced little solid evidence of its accuracy.[45]

In the absence of a thriving and benefit-delivering competition among private companies, many Denverites recorded on the editorial pages of local newspapers their conviction that municipal ownership was the better alternative. In 1874, the Colorado territorial government endorsed this line of reasoning. Alarmed by the wave of business failures, including more than a quarter of the nation's railroads, during the severe economic recession that Americans called the Panic of 1873, the territorial legislature wanted to ensure that an essential public service—supplying water to the territory's chief city—would not stumble during economic hard times. In 1874, the legislature amended the Denver City Charter to allow the city to "own water works of any description."[46]

The option of municipal ownership was a significant step up from the city's previous authority to grant franchises and regulate the city's water supply. Denver was now positioned in a productively adversarial role. It could grant a water franchise, but if the results were unsatisfactory, the city could build its own system. In response, the Denver City Water Company lobbied aggressively to be the city's water provider ("wining and feasting" the city council, according to one account) and succeeded in winning a seventeen-year franchise. While this was far from a municipally owned water system, the territorial legislature had provided Denver with a clearer and more powerful role in managing its water.[47]

The Denver Water Company (having dropped the word *City* in 1882) maintained its dominant position, but not without challenges. Denver's population grew through the 1880s, bringing new variations on the familiar problems of quality and quantity. After the death of president James Archer in 1882, the pace of company improvements slowed. Financial strain and internal disagreements over the direction of the company provoked squabbling among the remaining shareholders. A particular division of opinion centered on the question of whether the company should build larger storage facilities high up in the mountains.[48]

In 1889, Walter Cheesman and David Moffat, two of the original founders of the Denver City Water Company, stepped away from the internal feuding, sold their shares in the existing company, and formed the rival Citizens' Water Company. Gaining its own franchise in November 1889, their new enterprise competed directly with the Denver Water Company, bypassing local rivers and streams and heading straight for the mountains. The new company's managers wanted cleaner water and much more of it. They expected, as well, that the natural gravity flow offered by the drop in elevation would cut the expense of pumping. The founders of the Citizens' Water Company declared that they would "go direct to the mountains for a water supply that would permanently meet the city's increasing demands for quantity and quality." An enormous difference in the scale of this enterprise would put Moffat and Cheesman's company in an entirely distinct category.[49]

The first order of business for the Citizens' Water Company was to secure water rights for the South Platte River as it left the Platte Canyon. Purchasing agricultural land, with water rights attached, achieved this goal. Within a few months, the company completed a system of underground galleries in the mouth of the South Platte Canyon and finished the long pipeline that carried water to the city. An extensive network of pipes then distributed more than 8.4 million gallons of water a day to residents around the city. A belief in the superior quality of water acquired at a higher, more-isolated location was one of the key issues distinguishing Moffat and Cheesman from their former colleagues at the Denver Water Company and the primary reason they left the company. Happily for the two men and the people who drank their water, Citizens' product proved the validity of this belief: the water was remarkably clean and contaminant free. Denverites signed up for Citizens' services in droves, and by the end of 1890, the upstart company had captured one-third of the city's water market.[50]

In 1890, a drought descended on Colorado's Front Range. The Citizens' Water Company responded by expanding its system over the next two years, acquiring more water and building reservoirs. To bring more water into its system, the company added a second conduit from the South Platte, taking water from farther upstream and piping it to Denver. To enhance its storage abilities, Citizens' built the Ashland Avenue Reservoir and the Alameda Avenue Tank and purchased a large natural reservoir near Fort Logan, which it transformed into a storage reservoir called Marston Lake.[51]

Although it could not match the quality of water delivered by the Citizens' Water Company, the Denver Water Company ferociously defended its territory. The company swallowed its subsidiaries and merged with out-of-state interests, reincorporating itself as the Denver City Water Works along the way. Raising capital from East Coast investors and lowering its prices in an effort to persuade its customers to stay, the company fought its rival in any way it could. But Citizens' Water Company called its bluff and then went one step further, giving away water for free at various times in the early 1890s. The longed-for competition between Denver water providers was finally in play, but the contest was so fierce that it could not last long. Trying to match Citizens' provision of free water, the Denver City Water Works dropped all fees and thereby landed itself in receivership in February 1892.[52]

Triumphing in the water wars of the early 1890s, Citizens' Water Company purchased its rival's assets. With the two companies merged under a new name in 1894, the Denver Union Water Company was now the sole provider of water in the city of Denver.[53]

♦♦♦

The story of the companies that preceded the creation of the Denver Water Department, readers will have noticed, is a tangled tale. Its complications add up to vexation, and the storyteller tears her hair, seeking principles of selectivity that will allow her to omit tedious details without compromising the accuracy of the narrative. Readers, meanwhile, have reciprocal and proportionate reasons for tearing *their* hair, as they wonder how it is that western American history ever acquired a reputation as a lively, entertaining, and escapist subject.

Imagine if the planners of Disneyland and Disney World ever attempted to construct a theme park experience comparable to Frontierland but based on the romantic history of western urban water development. Increasingly desperate visitors would wander through a labyrinth in which they would encounter elite captains of competing and consolidating companies, complaining consumers disputing rates and service quality, city officials bending under these various inharmonic forces, occasional sufferers from typhoid, newspaper editors and reporters stirring up alarm over every new and old arrangement of authority, and an army of overactive attorneys. On the hour, all these parties would converge on the theme park's Main Street, wielding lawsuits, ballot initiatives, editorials, and company promotional material, for a picturesque struggle over the awarding of the next franchise, while tourists prepared the tales they would tell for decades about the worst vacation they ever took.

Would attention to corruption add sizzle and interest to this history and alleviate some of our storytelling woes? The Denver Union Water Company had its origins in the Gilded Age, an era in which businessmen's conduct dipped well below the ethical. Both David Moffat and Walter Cheesman, who would become the two most-significant figures in the early history of Denver's water supply, have been critiqued for their less-than-sterling performance in business ethics. Having thoroughly researched Moffat's career, historian Steven Mehls portrays Moffat as very much a participant in the customs of his times: "He often had a secondary regard for the safety of bank customers' money," Mehls observes. "He watered railroad stock, which was not unusual in that day, manipulated mining stock sales, was not always forthright with investors, and did all he could to advance his personal wealth." Mehls focuses his attention on Moffat's enterprises in railroads and mining, with only brief attention to the emergence of the Denver Union Water Company. But he gives no indication that the moral qualities of Moffat's operations in the world of water differed in any manner from his operations in railroads and mining.[54]

But was there anything particularly distinctive about the manipulations and exertions of power practiced by Denver's elite in the last half of the nineteenth century? All over the United States, variations on Denver's story unrolled in every major city. Citizens wanted new services, from water delivery to streetcar transport. The infrastructure required to provide those services could not be built without vast initial investments. The rationales and justifications for public ownership of this infrastructure, and for government spending of revenue raised through taxes or bonds to build it, were half-formed and far from robust. And so private enterprise found a great opportunity. It was hardly surprising that men who created these service-providing utilities expected a hearty return on their investments, and also expected to retain control of and authority over the systems they brought into operation. "To put it more clearly," a biographer of Walter Cheesman wrote, "Denver, during the first fifty years of its life, was a corporation built and controlled town and could not have advanced so rapidly because of its isolation had not this been the case."[55]

The boundary between public and private became most confounded by the practice of awarding franchises. City government granted franchises, giving individual companies exclusive claims on the provision of essential urban services. It would have been very remarkable indeed if franchises had been awarded purely on the basis of merit, experience, capability, and wisdom. On the contrary, the granting of franchises involved a great deal of behind-the-scenes jockeying, positioning, pressuring, and persuading. Once this process was under way, two boundaries often presented in public rhetoric as if they were stark and clear became notable for their fuzziness: the boundary between public institutions and private enterprise, and the boundary between honest effort and corruption. As Jane Haigh writes, "In Denver, as was typical in all US cities at the time, the City Council granted the opportunity to provide gas, electric, water, and transportation services to private utility companies . . . With so much to be gained from monopoly franchises for city utilities and transportation, the potential for corruption was enormous."[56]

Writing about the financing of the transcontinental railroads, historian Richard White has offered useful reflections on the topic of corruption. "Corruption is a species of fraud that involves violation of public or private trust," White declares. "Corruption involves betrayal, often of a third party. The corrupt buy or sell what was not supposed to be for sale—a vote, for example, or public property."[57]

The key phrase in White's analysis is *what was not supposed to be for sale*. In the last half of the nineteenth century, the definition of what was and was not for sale was dynamic and changeable. Americans in the late nineteenth century were, with variable success, trying to map out a boundary between what was for sale (the domain of the private) and what was not for sale (the domain of the public). The greatest riddle presenting itself was this: When the construction of an infrastructure served the needs of the general public, where was the line defining the level of legitimate profit for those who built that infrastructure and a scale of profit that crossed over the line into predation and even theft?

The most prominent builders of business empires in the Gilded Age, White notes, "reconciled morality and actions by embracing a morality of consequences." The distinction between bad and good was very clear in their minds: "Bad men lied and manipulated information to drive down values" while "good men lied and manipulated information to maintain or increase values." Using the methods they found necessary, good men helped investors and also helped cities and communities grow and develop, and thereby they displayed and affirmed their virtue. Rather than living with a sense of acknowledged, if also concealed, wrongdoing, "men of character," White concludes, "considered themselves the final judges of their own rectitude." Haigh applies the same insight to the principal figures of the Denver Union Water Company: "Moffat and Cheesman undoubtedly saw themselves as saviors and protectors of a locally run water system."[58]

In the move toward monopoly, the pattern of Denver's water development was also the national pattern in many enterprises: periods of fierce competition were followed by episodes of consolidation. Centralized power over essential services then inspired, in turn, the reform efforts of a wide range of activists who all fit in the category of antimonopolists. Alarmed by concentrated power in transportation, mining, manufacturing, food processing, finance, banking, and utilities, antimonopolists, in the manner of fighting fire with fire, found a new enthusiasm for the centralized power of government. It is, after all, one of the striking aspects of this tangled tale, even in an era with a great commitment to privatization, that the idea of government ownership was in play so early in the game. The purchase of the Big Ditch and its rechristening as City Ditch provided one example, as did the early pleas from citizens and newspaper editors to rescue residents from dependence on powerful companies with municipal ownership.

"Western promoters," historian David Wrobel wrote in his study of boosters, "hurriedly raced toward the future, often announcing its presence before it had actually arrived." Boosters were "optimistic fortune-tellers who told present and prospective residents what they wanted and needed to hear about western places. They placed the clear, bright future in the cloudier, less certain present."[59]

Denver's boosters exemplified this mind-set, but they were also, in comparison to a more flighty and aerial variety of booster, intensely practical and down-to-earth in making investments and taking action to bring their vision of the future into existence. One writer aptly described Walter Cheesman and his associates as a "small but extraordinary group of strong men" who "envisioned Denver" as one of the "great metropolitan centers of America" and

From left to right, Denver Union Water Company officials David Moffat, William Robinson, and Walter Cheesman pose for a picture in January 1905. Moffat and Cheesman were successful, civic-minded businessmen who founded the Denver City Water Company in 1870 and subsequently the Citizen's Water Company in 1899, forerunners of the Denver Union Water Company and ultimately today's Denver Water. All three men served as president of Denver Union, a position to which Robinson succeeded them in 1914.

"as the focal point of a western empire." "No doubt, we would consider these dreams extravagant today," wrote Edgar C. McMechan in the 1930s, "but the leaders of that day backed their belief with indomitable wills."[60]

Similarly, Jane Haigh characterizes the small group centered on Moffat and Cheesman as "elite businessmen who believed that what was good for them and their businesses was good for Denver and for Colorado." The scale of profit they expected as a return on their efforts, as well as the methods they sometimes used to press their own interests and squash opposition, can certainly be subjected to a hindsight critique, particularly in a latter-day era when population growth has become, for many, a subject of lament. And yet the assumption—that progress for the utility entrepreneurs also meant progress for many others—carried unmistakable dimensions of accuracy: many people found homes and opportunities in Denver because Moffat and Cheesman had pursued their goals without restraint. As Steven Mehls puts it, Moffat stayed "busy with projects designed to convert the rough frontier town of Denver into a true urban environment" and "enticed millions of dollars of outside capital into Colorado and in so doing helped the state's rapid economic growth from 1860 to 1911." In a recent poll, Coloradans reported a striking, even flabbergasting level of happiness. As political scientist Thomas E. Cronin summarized their cheer, "A whopping 97 percent of adult Coloradans polled say they are 'generally a happy person.'" Should these folks try to live in the conditions described by the Pike and Long expeditions, the percentage of the happy would certainly shrink. Directly and indirectly, the undertakings of Moffat, Cheesman, and their associates add up to a significant force in laying the material foundation for widespread well-being.[61]

Though "situated in what is not improperly termed an arid region," Denver "has become the financial, commercial, and manufacturing metropolis of the vast empire of arid country," Jerome Smiley wrote in 1901. "It could not have become such nor could it continue greater with a water supply insufficient for the daily needs of the people and industry within its limits."[62]

With 106,713 residents in 1890, two telephone companies, streetcars running to suburbs, and daily rail shipments of goods produced in remote locales, Denver differed dramatically from the mining town founded barely thirty years earlier. Far from the "treeless, grassless, bushless" settlement with seemingly improbable ambitions that it had been, Denver in 1890 resembled its urban counterparts in the eastern United States. In a remarkably successful defiance of the climate, the irrigated Eden prophesied by early boosters had moved from imagination to reality.

"No other community so large as Denver is so peculiarly circumstanced with relation to its water supply," Smiley wrote in a statement that did not exactly pay due respect to Los Angeles or San Francisco. "In other cities the great problem is to efficiently distribute to consumers water readily accessible and practically unlimited in quantity," he explained; "here the problem is—or was—to secure water for distribution."[63]

More than a century later, Smiley's statement is a valuable reminder of just how improbable the development of Denver was. The engineering achievements of the nineteenth and twentieth centuries transformed American life so completely that they erased from view the limitations that natural conditions once exerted over settlement.

Equally important was Smiley's shiftiness in verbs: Denver's "problem is—or was—to secure water." *Is* or *was*? Did the problem still exist, or had it been resolved with the vigor of the Denver Union Water Company? Or had the problem of water's scarcity simply been postponed and handed off to the next generation? Jerome Smiley was writing in eager anticipation of the Denver Union Water Company's Cheesman Dam. Even before the dam was completed, in what would become a long-lasting pattern of expression, its supporters had already taken to declaring that the completion of the next big project was going to solve the region's water problems. The engineers, meanwhile, would put their ingenuity to work on providing the supply that would support the next big population surge and thereby bring on the next confrontation with overstretched resources. Over the next century, prophecy and ingenuity would hand the city's destiny back and forth—the baton in a dramatic and often nerve-racking relay race.

A Failure to Communicate

Though we would never want to be catty,
Engineers have made nature ratty.
As they fill all our needs,
We bombard them with screeds,
Which for some reason drives them all batty.

Chapter Two

Go Take It from the Mountain

The modern technologies of sanitation [and water supply] came to be regarded as one of the major success stories of city building in the Progressive Era, systems that would stand the test of time and provide a solid basis upon which to build thriving communities.

—Martin V. Melosi, *The Sanitary City*[1]

In the turn-of-the-century American West, the down-to-earth power of ideas was unmistakable. In concrete terms (literally), thoughts, visions, and expectations rearranged the landscapes and rivers, as well as local configurations of political and economic power. With the construction of a great mountain reservoir, Denver put the idea of defying nature's limits into very direct action. To the city's leading figures, the canyons of the Rockies presented themselves as reservoirs waiting to happen. Over the course of a century, this vision produced an elaborate network of dams, reservoirs, tunnels, ditches, pipes, pumps, and filtering plants that reconfigured the arrangements of water and made possible the creation of a big city in an improbable setting.

While this story has some elements that are distinctive to Denver, the same basic tale unrolled in many western locations. The historian Carl Abbott describes the broad pattern: "Residents of western cities have reached far into their hinterlands and remade distant landscapes in their own interests." All around the West, Abbott explains, the leaders of ambitious cities "had the ambition and financial capital to reach deep into high mountains with increasingly elaborate engineering and to divert water over natural barriers."[2]

In the same era, a parallel process of political construction took place, as cities moved water systems from the ownership of private companies to municipal organizations funded by public money and run by public officials. Both the construction of dams and the creation of agencies demonstrated a familiar historical pattern: when ideas go into practice, they can generate as many unintended consequences as desired results.

In the last decade of the nineteenth century and the first two decades of the twentieth century, western cities expanded their water-supply systems and changed the structures of governance of those systems. In these actions, the cities displayed a similarity of strategy that almost seems coordinated and orchestrated (there are marching bands and synchronized swimming teams in the world that have seldom achieved a similar level of precision). In every locale, the conviction that prosperity required an abundance of water provoked these parallel actions. Even if supply momentarily satisfied demand, a glance at the future left no use for the words *adequate* or *sufficient*.

To many well-informed westerners today, the two celebrities in the story of urban water development are the cities of San Francisco and Los Angeles. More than a century ago, San Francisco developed a desire for the Tuolumne River in the Hetch Hetchy Valley, and Los Angeles targeted the Owens River. These two historical episodes have become legends, parables, and cautionary tales. Meanwhile, the historical episode in which the city of Denver built the massive Cheesman Dam in the canyon of the South Platte River earned far less fame, visibility, or notoriety—an inequitable distribution of attention that requires explanation.

Controversy provides the first clue. Because they met significant resistance and became subjects of national dispute, San Francisco's dam in the Hetch Hetchy Valley and Los Angeles's aqueduct from the Owens Valley lodged in memory and became familiar points of reference in public debates of the twentieth and twenty-first centuries. Denver's Cheesman Dam, by contrast, never attained any such fame.

In our times, it strains the imagination to think of a noncontroversial dam. But in the past, it was not at all uncommon for dams to come into existence without controversy and even with widespread approval. In that sense, the water projects of San Francisco and Los Angeles were anomalies and exceptions. To put this more accurately, the fights they triggered were unusual early examples of what later would become a common pattern. In the allocation of controversy, both California cities faced two big factors that were conspicuous by their absence in Denver's situation.

The canyon that the Denver Union Water Company chose for Cheesman Dam and Reservoir did not have a population of enthusiasts, devotees, or nearby residents attached to it. The canyon in the South Platte did not hold a settlement of farmers, ranchers, bankers, merchants, and attorneys, as the Owens Valley did. Nor did it come with a set of defenders equivalent to the upper-class nature lovers and recreationists who fought for Hetch Hetchy. In comparison to San Francisco and Los Angeles, Denver's developers had extraordinary freedom to plan and to act without opposition from articulate and impassioned fellow citizens. The members of the Sierra Club and the residents of the Owens Valley succeeded in casting the cities of San Francisco and Los Angeles as imperial antagonists and shameless bullies and thus shaped the image of both cities for decades to come. There was no comparable group mobilized to denounce and oppose Denver and the building of Cheesman Dam.

The other great difference in Denver's circumstances lay in the largely perfunctory part played by the federal government in the story of Cheesman Dam. Since Hetch Hetchy lay within the boundaries of Yosemite National Park, San Francisco first needed permission from the secretary of the interior to put a dam and reservoir within the park's borders. Over the course of a decade, secretaries of the interior came and went, and the status of San Francisco's permission to proceed fluctuated with these comings and goings. After various reversals of decisions, one secretary declared that the power to make the decision did not, after all, lie with the Department of the Interior and required instead an act of Congress. At that point, the city's officials had to relocate to Washington for a prolonged spell of making their case to the members of the House and Senate, lobbying for a bill to award the city the authority to build in a national park. All this kept the project front and center in national attention.

In a comparable way, the fate of Los Angeles's plans rested on acquiring the right-of-way to build the aqueduct across lands in federal ownership. Like San Francisco, Los Angeles had to make its case to Congress in order to secure that right-of-way. The Owens Valley project, moreover, had its origins in a tangled relationship with the new Reclamation Service, with a key figure employed by Reclamation working as a paid consultant for the City of Los Angeles, a situation that left the project with a lasting association with impropriety and conflict of interest. Struggles between city ambitions and federal authority not only added to the controversy, but also added years to the time before the projects could be started and completed.[3]

The federal government's role in Denver's early plans was dramatically less controversial. The South Platte Canal and Reservoir Company (the construction wing of the Denver Union Water Company) filed for the necessary rights-of-way over federal land in 1894. The secretary of the interior approved a permit for a five-year construction period in 1895. The company returned and asked for a five-year extension, which it received in July 1900. In 1902, the company completed its purchase of the reservoir site from the United States at $1.25 an acre, after "this purchase was authorized by an act of Congress, originally sponsored by [Colorado] Senator Henry M. Teller," with the site augmented by the purchase of eight hundred acres of state school land. If congressmen had any doubts about allowing the land to be purchased for a dam, they could find comfort in the appraisal of Lieutenant Colonel Hiram Chittenden of the US Army Corps of Engineers, who confirmed Denver Union's judgment that the site was "an excellent place for a high masonry dam" in his 1898 survey of the area.[4]

Compared to the tangled chain of events leading to the Hetch Hetchy Dam and the Owens Valley Aqueduct, in which municipal agencies were driving the development of water systems, the federal dealings that led to Denver's Cheesman Dam project were simple and uneventful. In Denver's case, a private company sought to use federal resources, a much more familiar situation with many precedents behind it than the comparatively new and uncharted territory of relations between a city agency and the federal government.

Lucky Denver! Thanks to an early start in the mid-1890s, the absence of outspoken opposition, the speed and smoothness of federal decision making, and the nimbleness possible for a private company in contrast to a city agency, Denver had a functioning dam and delivery system in place by 1905. Agents representing Los Angeles began eyeing the Owens River in 1904; the system did not deliver water to the city until 1913, putting Los Angeles notably behind Denver. San Francisco first began considering the Hetch Hetchy Valley for a dam in the late 1890s, but construction did not begin until 1913. For various reasons, including complicated dealings with a private water provider, the Spring Valley Water Company, San Francisco did not begin using water from Hetch Hetchy until 1934.

Thus, rather than affirming the conventional image of California as the pacesetter and leader of change in the American West, the chronology of urban water development actually assigns the role of pacesetter to Denver, leaving Los Angeles and San Francisco trailing behind. And yet leaders of Los Angeles and San Francisco came out of their travails well ahead of leaders in Denver when it came to an early recognition of the complications, difficulties, and challenges of building big water projects. A later generation of Denverites would eventually catch up with their California urban counterparts when it came to facing forceful opposition, running a gauntlet of federal decision making, accumulating a reputation as a despoiler of pristine rural places, and ending up with a much-delayed project or no project at all.

If Denver set records for speediness in building its first big dam and reservoir, its pace slowed considerably when it came to constructing the organization that would manage the water system in which Cheesman Dam was the principal feature. To those who would be hard put to imagine, much less to undertake, a strenuous, complex, and detailed task like the building of a big dam, it might seem that building a new city agency would be by far the easier task.

Guess again.

The transitions—from a competition among private water companies, to a monopoly under one company, and then to the acquisition of that company by the city—took a lot longer and involved many more twists and turns, advances and reversals, than the building of Cheesman Dam. The Colorado Territorial Legislature gave incorporated cities the power to grant franchises to water companies in 1872. The Denver City Charter of 1874 gave the city the authority to "own water works of any description."[5]

Despite this early empowerment, the city did not complete its purchase of the Denver Union Water Company until 1918. Even with a very big setback from a flood in the middle of construction, Cheesman Dam took less than a decade from the start of construction to the delivery of water to the city. It took the better part of half a century to bring the Denver Water Department into being. Coordinating thousands of voters to approve a plan for and the funding of a municipal water system took decades, while coordinating hundreds of workers to operate as a team to build a dam took a few years. People familiar with the challenges of democratic decision making should be well-equipped to guess at some of the reasons behind this very different pacing.

♦♦♦

In 1894, after thirty years of competition and consolidation among companies, the Denver Union Water Company had emerged as the winner, launching a new era in which a single company would supply water to the entire city. The familiar phrase "to the victor go the spoils" had some bearing on the situation, but "to the victor go the expressions of citizen discontent, dissatisfaction, dismay, and disappointment" had nearly equal relevance. And yet its dominant position also gave Denver Union the power to act on the plan to secure an ample and pure supply of water by building an enormous reservoir high in the Rocky Mountains on a scale that neither the city—nor much of the world—had seen before.

By 1900, the city's population exceeded 130,000. The local manifestation of the City Beautiful movement had rewritten Denver, declaring the town's prosperity in a vocabulary of verdant parks, substantial buildings, and tree-lined esplanades. The increase of population and the comfort and health of the residents seemed to rest on the maintenance and expansion of a clean, predictable water supply. Surrendering the nineteenth century's faith in providential changes in climate and weather, Denver's leaders shifted their faith to the capacity of engineers to correct nature's deficiencies.

In city after city, the reckoning with future desire for water presented a spectacular opportunity to the members of the emerging profession of engineering. Advances in hydrology and hydraulic engineering in the late nineteenth and early twentieth centuries dramatically increased the ability of engineers to estimate annual precipitation and storm frequency, capture the water, and deliver it on demand to people far from the source. Many engineers saw themselves, and were seen by many citizens, as forces of reason, taking cities that had grown chaotically and using efficiency and foresight to tame them. As historian of technology Carroll Pursell puts it, "every city had an extensive (and expensive) engineering infrastructure of streets, sewers, water supply, and electrical lines that formed the skeleton and nerves of the urban body," and engineers were the crucial figures in designing and installing this essential infrastructure. In line with historian Edwin Layton's characterization of them "as stewards of technology," engineers saw themselves as "the agents of social progress." Indeed, in 1913 the president of the American Society of Civil Engineers, George F. Swain, would proclaim that the engineer was "possibly better fitted than anyone else" to provide answers to the social problems of the day.[6]

Denver Union's engineers were thus representatives of a much larger national and international deployment of human ingenuity and resolve. Their task involved the double aspects of improving the quality and increasing the quantity of the water the company provided to the residents. The company created a laboratory to keep track of the purity of the water, experimenting with chlorination as a means to guard against waterborne diseases like typhoid and cholera, and enlarged the Marston Lake Reservoir, southwest of the city, to increase the supply on hand. Projects like the Marston expansion helped the company keep up with the city's growth, but these were incremental, stopgap methods. David Moffat, Walter Cheesman, and their colleagues—first at Citizens' Water Company and then at Denver Union Water Company—had cast their ambitions on a much larger scale.

In 1891, Citizens' Water Company had proposed the Strong Reservoir project, which called for a pair of immense rock-filled dams, 275 feet high and more than 1,000 feet thick. About thirty miles southwest of Denver, near the town of Morrison, these two dams would hold water from Little Deer Creek and Brush Creek in two massive natural basins. The extraordinary scale of the project attracted nationwide attention in popular publications like *Frank Leslie's Illustrated Weekly*. With striking drawings and heightened rhetoric, the magazine declared that the imagined project would be "the greatest gravity system in the world," celebrating the care that Citizens' leaders had taken to choose a site where gravity would make expensive pumping unnecessary.[7]

Such an immense project was necessary, as James D. Schuyler, a distinguished hydrological engineer and a consultant to the company, wrote in a report to the company's board of directors, "to provide an ample water supply, not only for the present population of the City, but to lay the foundation for a supply which shall be adequate for the wants of a City of 500,000 inhabitants, which your faith in the future of Denver leads you to believe it will attain within the next fifteen to twenty years."[8]

Schuyler's remark was a classic demonstration of the thought process that would drive urban water development: pick a date in the future; make a guess at the number of residents likely to call the city home in that year; think of this growth in numbers as right and necessary; and express the certainty that their demands for water would be a matter of wants or needs, and therefore not up for negotiation. Although the company finally dropped the Strong Reservoir plan, the proposal had demonstrated the extraordinary scale and unprecedented effort of the water development plans of the dawning twentieth century.

Visions of the Strong Reservoir yielded to a plan for a dam at a site with much more to recommend it. Citizens' Water Company chief engineer C. P. Allen first scouted this superior site on September 23, 1893, while combining business and pleasure on a surveying excursion and fishing trip on the South Platte River. Nearly fifty miles southwest of the city, just below where Goose Creek joined the South Platte, towering granite cliffs converged in on the river, creating a sliver of a gorge only thirty feet across at its base, widening slightly to a (still narrow) one hundred and thirty feet. This sculpted space meant that building the dam would require less in the way of material, labor, and expense. In the Platte Canyon, nature had stepped in and done a significant share of the construction work. And, while the earlier plans for Strong Reservoir had predicted a 9.5 billion-gallon capacity, the smaller but stronger dam built in the deep and narrow Platte Canyon would eventually hold back more than 25 billion gallons of water. The dream of a high mountain reservoir, sited far above the contaminations and pollutions of human settlement and storing a vast supply, finally seemed within reach for the people of Denver.[9]

When construction on the dam began in the spring of 1899, the plan called for a large gravity design rising two hundred feet above the riverbed. The physics of a gravity dam required an enormous amount of material to withstand the horizontal force of the water: thick steel plates (dipped in asphalt to keep the dam watertight) riveted onto steel beams would loom upward from the bedrock and form the core of the dam; solid granite block masonry walls, methodically assembled in place, would flank these plates to provide stability; and, behind the dam, a layer of concrete and tons of earth and rock-fill—enough to bring the dam's total thickness to more than six hundred feet at the base—would be piled to hold the water at bay. To ensure its structural integrity, the dam had no breaches through the masonry. Instead, workers bored large outlet tunnels through one hundred twenty feet of granite to divert any excess pent-up water through the solid walls of the canyon. Once the reservoir was filled, valves in the tunnels would control the purposeful discharge of water. The dam was intended to function

Three engineers stand in the narrow space between the walls of Platte Canyon in the 1890s, surveying the spot where the Denver Union Water Company would erect Cheesman Dam.

neither as a flood-control mechanism nor as a significant source for hydroelectric power, but simply to store a reserve of pure water for year-round use by the city's residents.[10]

The site's rugged terrain challenged laborers and engineers alike. Workers had to negotiate the precipitous canyon wall and perform their jobs on winding wooden stepladders, often in snow in the winter. The project's designers and managers struggled to devise methods for supplying material from sites located in the same kind of steep, harsh terrain as the dam itself. In one example of their necessarily innovative approach, they created a quarry atop the adjacent cliff and transported stones and rock-fill from it via a custom-built railway that spanned the canyon directly above the dam, allowing workers to dump fill material directly into the gorge below. As this example and many others demonstrate, in order to build a dam, workers and managers had first to build a whole other infrastructure to supply and support the construction.[11]

Although the Platte Canyon site was ideal for a dam, the rugged terrain challenged the builders. This view, taken in May 1899, shows precarious wooden stepladders on which workers negotiated the steep walls and the false work for a bridge across the chasm from which crews would deposit fill material behind the dam.

Despite the difficult circumstances, construction moved forward, and by April 1900, the steel and masonry wall had reached fifty feet, and the rock-fill behind the dam was even higher. Denver seemed on the verge of securing its pure, mountain reservoir. But a snowy winter and high spring runoff meant a rapid rise in the water behind the dam. The recently installed main outlet valve was opened as wide as it could go, and workers frantically piled earth and rock atop the dam. But on the morning of May 3, 1900, the water surged over the masonry wall.[12]

With impressive and lifesaving foresight, chief engineer C. P. Allen quickly organized teams of Denver Union employees to fan out, on foot and on horseback, to warn downstream communities of the approaching flood. William Flick, foreman of a railway grading crew that was working below the dam, described the alarm system: "Paul Amos galloped down the canon [*sic*] like wildfire, shouting that the dam had broken and for us to get out of the way! . . . He was going like a flash, but the wave was dead after him and it must have been a fine race." According to company lore, Denver Union construction worker Zebulon Swan also leaped onto a horse

Although the Platte Canyon site was ideal for a dam, the rugged terrain challenged the builders. This view, taken in May 1899, shows precarious wooden stepladders on which workers negotiated the steep walls and the dam's wooden false work for a bridge across the chasm from which crews would deposit fill material behind the dam.

Despite the difficult working conditions and harsh winter weather, the construction crews made steady progress. This view shows builders erecting the masonry wall in January 1900. By April of that year, the wall had reached fifty feet, and the rock-fill behind the dam was even higher.

and tore off down the canyon ahead of the water, warning everyone he encountered of the torrent to come. As his descendants, themselves Denver Water employees, relate the family legend, Swan rode the horse to its death in the effort.[13]

The people of downriver towns were grateful for the efforts of men like Allen, Amos, and Swan. The *Rocky Mountain News* reported that the residents of Littleton, one of the first towns along the river as it came out of the mountains, "were thoroughly alarmed about 10 o'clock in the morning by the runners and couriers of the water company, who warned everybody in the low lands to get out before noon, as the flood would surely arrive at 1 o'clock. The rise came on time."[14]

For those watching from a safe distance, the devastating power of the deluge awed witnesses. William Flick described the destructive power of the flood as it crashed down the canyon:

> I never saw such a sight. We saw the wave half a mile before it got to us. It was a solid jam of logs and debris. Everything was clawed into the river. . . . Just below us, in a narrow place, the wave was fully fourteen feet deep, and traveling like a train. Our camp went up like a rocket. . . . An iron bridge this side of South Platte was twisted up like so much line wire. Above the South Platte the water washed out the grade and left a granite wall thirty feet high standing there. The entire wagon road of twenty-three miles is gone, and one-third of the track from South Platte to Platte canon [*sic*] has been washed out, rails and all. Telegraph poles are likewise cut off like pipe stems.[15]

Back in the canyon, the company's small railway bridge above the dam now provided a view not of the Platte Canyon dam, but of the loss of two years' worth of labor.

💧💧💧

The response to this calamity offers quite a dramatic contrast between the attitudes and practices of the early twentieth century and the attitudes and practices of the early twenty-first century. Such a catastrophe, in our time, would surely keep everyone in sight preoccupied with lawsuits and liability, recriminations and second thoughts, regret and reproach, for a decade or two. In 1900,

After the flood, the view in the canyon showed the loss of two years' work. This photo shows water flowing sixteen inches deep over the masonry wall on May 8, 1900.

the washout of the dam was certainly a disappointing setback. But within two months, the company's board of directors was ready to try again. The new structure, rising on the foundation of the first, was both more ambitious and less risky.

After reviewing several competing proposals, Denver Union chose the innovative design of globe-trotting engineer Charles L. Harrison, inspired by his promise that the new dam would be "so reinforced and anchored that no flood could disturb it." Harrison devised an elegant and radically different plan for the second attempt at damming Platte Canyon: the gravity arch. A pure gravity design, as the first dam had been, relies on the sheer weight of the rock to fortify the dam. An arch dam draws strength from its shape, with solid granite walls that arch like a tightly drawn bow while the weight of the water pushes the dam's stones firmly together. Harrison combined elements of both designs into a gravity arch hybrid that curves toward the reservoir at a 400-foot radius on the upriver side, while the granite stones in the rear taper down like a giant staircase 18 feet thick at the top to 176 feet at the base.[16]

Working conditions were no easier the second time around. To get to the site, laborers, equipment, and supplies first traveled to a railroad depot twenty-three miles from the remote canyon. From there, horse-drawn wagons and, later, a company-built railroad hauled men and machinery over the beautiful but rough mountain terrain to and around the dam site. Workers lived in shacks arranged around the small scattering of support buildings constructed for the duration of construction. These included a quarry office, a blacksmith shop, a cook shack, and even a tiny post office. Work on the dam continued year-round, even through the Colorado winters, when roughshod horses pulling sleds replaced barges carrying heavy construction materials—including 27 million cubic feet of masonry and 100,000 barrels of cement.[17]

Completed in 1905, the 221-foot Cheesman Dam became the highest dam in the world, a status it would hold until 1912.[18]

♦♦♦

In 1860, territorial governor William Gilpin declared that the region centered on Denver would become a pastoral garden. By the first decade of the twentieth century, with Cheesman Dam as the centerpiece of a system that brought water compliantly from the mountains to the metropolis, water was put to work to validate and vindicate Gilpin's prophecy. It filled the bathtubs and drinking glasses of Denver's residents and showered the lawns around the gold-domed state capitol building. Guided by expectations formed and confirmed in places with rainfall far greater than Colorado's, the city's exuberant landscaping was a point of civic pride, from the flower gardens of private homes to the artful plantings of the city's new parks. City employees regularly sprinkled the unpaved streets to keep dust down. "Ugly things do not please," explained Denver's mayor Robert Speer (1904–1912, 1916–1918). "It is much easier to love a thing of beauty—and this applies to cities. Fountains, statues, lights, music and parks make people love the place where they live." Many of the city's new parks included water features as their central attraction, from the waterfall and pools of Sunken Gardens Park to the fountains in City Park. Beauty, in an assumption that Speer shared with many of his fellow citizens, was at least damp, if not entirely soaked and saturated.[19]

When Cheesman Dam was completed, both Denver Union and the citizens of Denver saw it as the solution to the city's water problems. But by the 1920s, Denver's rapid growth would call such certainty into question, requiring another undertaking in the series of "permanent" solutions to its problems. By enabling the city to expand without immediate constraint, Cheesman Dam established the conditions that would eventually render it inadequate. For a period of time, the abundant water supply provided by the dam encouraged a growth in population that would finally exceed the reservoir's capacity to satisfy the demands placed upon it. Such a shortfall would inspire a succession of ambitious water projects that repeated this pattern of an increase in water supply, making possible an increase in population growth and creating a demand, in turn, for the next increase in water supply.

Today one can visit bigger, more impressive dams built later in the twentieth century, but Cheesman retains a unique place in American history. More than a century after its creation, the dam still has some of the elements and features usually associated with works of art. The stones, precisely cut by expert masons from native granite, merge seamlessly into the cliffs on either side. The spillway, designed by Alexander Kastl, sends water over the north side of the dam, cascading down the cliff in a very exact imitation of a natural waterfall. The reservoir itself could nearly pass for a natural lake, as small beaches have formed where decades of lapping have worked over surrounding boulders. Its enduring functionality—and its lasting elegance—testify to the innovative engineering that got Denver its water. In 1973, the American Society of Civil Engineers recognized Cheesman Dam's significance to Denver and to the engineering profession by designating the dam as a national Historic Civil Engineering Landmark.[20]

When you stand on the downstream side of Cheesman Dam and look up to see the way the structure joins with the canyon walls, with curve and variation everywhere, the beauty is so apparent and so compelling that a heretical thought shoots into the mind: the National Endowment for the Arts should award a counterpart comparable to its designation as Civil Engineering Landmark, a National Landmark in the category of Land or Environmental Art. At the risk of sailing off (temporarily) into the perplexing seas of academic language, consider art critic Jeffrey Kastner's definition of this genre:

> The range of art work referred to as Land Art and Environmental Art . . . includes site-specific cultural projects that utilize the materials of the environment to create new forms or to adjust our impressions of the panorama: programmes that import new, unnatural objects in to the nature setting.

So far, Cheesman Dam seems to be meeting the admission requirements for the genre, a judgment only furthered by Kastner's observation that, in its origins, land art was "the most macho" of art forms: "In its first manifestations, the genre was one of diesel and dust, populated by hard-hat-minded men, finding their identities away from the comforts of the cultural centre, digging holes and blasting cuts through cliff sides, recasting the land with 'masculine' disregard for the longer term." It is hard to imagine how the dam could do more to qualify for a high rank by this particular measure.[21]

If the Denver Water Department, in the early twenty-first century, submitted Cheesman Dam for inclusion in a juried show of land art, its very evident qualifications for acceptance would, nonetheless, be derailed by its three fatal flaws: its timing is all wrong; it is compromised and tainted by its usefulness; and it was designed and put in place by engineers. The art historians are, after all, very clear on both the identities of the participants and on the chronology of this genre. Land art is a post–World War II movement that

Completed in 1905 and still a vital piece of the Denver Water supply system, Cheesman Dam stores nearly 80,000 acre-feet of water for Denver and serves as an example of engineering as land art.

"began in the mid 1960s with a small number of committed conceptualists" who were "disenchanted with the modernist endgame." The engineers of the Denver Union Water Company had clearly jumped the gun in this race toward artistic innovation![22]

To the interpreters and analysts of the genre of land art, Cheesman Dam must be sent back to engineering and denied entrance to the world of art. It may pass itself off as an object of beauty and an achievement of human creativity, but it is an industrial intrusion, a utilitarian device, and a major disruption of a free-flowing river. But to the rest of us, unsophisticated and uncredentialed as critics, an intense and open-minded contemplation of Cheesman Dam leaves our categories and concepts productively blurred and even scrambled. We may have to return from that episode of contemplation with a recaptured regret and dismay over the intrusion of human meddling in natural systems. But we will also return with the honest recognition that many of the artifacts produced by that meddling carry an unsettling beauty of their own.*

Thanks to the artful Cheesman Dam, water flowed plentifully all through Denver, and most residents made correspondingly liberal use of it. Denverites enjoyed a "lavish, almost unrestricted, use of water," remarked Jerome Smiley at the turn of the century.[23]

The Denver Union Water Company made this abundance. As the city continued its rapid growth in the twentieth century, Denver Union continued to design new ways to give the citizens of Denver the water they had come to expect. Denver Union's ascent was rapid and remarkable, but maintaining its dominance could turn out to be as tricky as achieving it in the first place.

♦♦♦

As the cities of the early twentieth-century West expanded, they often grew so quickly that they outpaced their municipal utilities and supply systems. Responding to the demands of their growing city, Cheesman and Moffat had created a monopoly that, in the minds of its critics, held far too much power over the well-being of

---

* My thoughts here have some connection to what David E. Nye and other scholars have called "the technological sublime." And yet what I have in mind is a much more twenty-first-century sentiment in which the viewer is well aware of the costs and burdens of the admired human artifact, but nonetheless is swept away by the creativity and artfulness it embodies. See David E. Nye, *American Technological Sublime* (Cambridge: MIT Press, 1999).

As befits such a prominent figure in Denver's water history, Walter Cheesman's Capitol Hill mansion at the corner of Logan Street and Eighth Avenue was surrounded by trees and lush lawns. Cheesman died before construction was completed in 1908, but his wife lived in the home for more than a decade. In 1959, it was donated by the Boettcher family to the State of Colorado as the governor's official residence, a role it still serves today.

the city. Like many Americans unsettled by the expanding powers of corporations and businesses, a significant sector of Denver residents distrusted the company and argued for outright public ownership.

The opponents of the Denver Union Water Company held a place in the most important political trend of the late nineteenth and early twentieth centuries: the growing force of antimonopolists and their campaign against "aggregations of capital." As historian Richard White puts it, "A whole array of people in the Democratic, Republican, Populist and Progressive Parties thought that monopolies or corporations, two words they used almost interchangeably, had acquired special treatment and privileges from the government (tariff protection, subsidies, the use of troops against their opponents, special legislation)." These government-bestowed privileges had produced such a very alarming state of affairs that "ordinary citizens could no longer think that success depended on competition or fairness."[24]

Taking up causes beyond number and measure, the antimonopolists searched for remedies to this disturbing situation. Full participants in this broad national campaign, Denver residents and city officials pitched into a series of disputes between the city council and Denver Union over water quality and rates, increasingly making the case that the domestic water supply would be better managed by a publicly owned enterprise. This was not, after all, a new idea; as early as 1866, the territorial legislature had given the city the authority to provide water for irrigation, and the city had, over the next years, purchased the Big Ditch and renamed it the City Ditch (see chapter 1). In 1874, the territorial legislature further authorized the city to "own water works of any description." And yet the acquisition of the Big Ditch was a very limited precedent in the category of owning waterworks; compared to water for outdoor irrigation, providing water pure enough for domestic use required much greater investment and responsibility. An even greater obstacle to municipal ownership lay in the question of setting a fair price for the facilities already created by private companies. Throughout Denver Union's existence, the city government made several attempts to purchase Denver Union's entire system, but conflict over the price interrupted each of those efforts. At one critical point, when the possibility of buying the system seemed bleak, municipal leaders even considered building their own water system from scratch. But the cost was prohibitively steep.

Public utility versus private utility: this debate took place in many cities, with great animation, intensity, and often anger. As historian Martin Melosi wrote, "Water became a particularly favorite political issue because embedded in it were so many concerns touching the well-being of the citizenry as well as the role of government in serving that citizenry." The West's distinctive practice of allocation—the famed doctrine of prior appropriation—gave the debate a distinctive regional cast. The ability to secure a senior claim to water—wherever it might be originally flowing—and to then transport it far from its point of origin allowed Los Angeles, San Francisco, and Denver to seize the resources of their hinterlands and direct them to the support of metropolitan growth. Established in the last half of the nineteenth century, the doctrine of prior appropriation was well-suited to a society that valued water primarily for its power to support profitable economic activity. Prior appropriation provided the essential mechanism for extending the reach of cities into their hinterlands and thus for raising the stakes of municipal power.[25]

In 1872, the territorial legislature had given cities the power to award franchises to water companies, giving them an exclusive right to furnish water to particular areas. In 1874, prompted by the legislature's decision to give Denver the option to own a municipal water system, the Denver City Water Company sought and won a seventeen-year franchise to provide water to the city's taps, a relationship the city government renewed in 1890 for an additional twenty years. In 1894, after a swirl of competition among companies, when the last two survivors—the Denver Water Company (having dropped *City* from its name in the interim) and the Citizens' Water Company—merged into the Denver Union Water Company, the two companies brought to the relationship two separate, preexisting franchises. The twenty-year franchise awarded to the Denver Water Company in 1890 gave more power to the company and less power to the city than did the franchise given to the Citizens' Water Company in 1889. Predictably, when the Denver Union Water Company emerged as the sole provider for the city, its owners chose to hold on to the more favorable twenty-year franchise. By virtue of the very same quality (the advantages that the franchise gave to the company), in the judgment of many citizens, the terms did not give the city and county governments sufficient power to require affordable rates and a higher quality of service.

But the 1890 franchise did provide that at the end of its twenty-year period, "the said works may be purchased by the city."[26]

In other words, everything necessary was in place for a long and draining tussle over the new company's monopoly over the essential resource of water, a resource that provided the foundation for every aspect of the city's well-being. Following a well-worn course established with the earlier Platte Water Company and Denver City Water Company, Denver Union's customers soon complained about the company's water rates and water quality, making sure that city officials knew of their discontent. Calls for a municipally owned water system grew in number and in volume. Water was a principal issue in the mayoral election of 1897; when Mayor Thomas S. McMurray vowed to fight the water company, the declaration helped earn him another term in office. Later that year, the city council sued the water company for not living up to the provisions of the franchise, citing excessive rates, low water pressure, and poor water quality. It would take ten years for the courts to settle this case. In the meantime, the city tried repeatedly to force the company to lower its rates.[27]

The struggle had reached no clear conclusion. Public complaints continued, leading some city officials to resolve either to purchase Denver Union's assets or build a separate system. In May 1898, Denver Union offered to sell all its property, water system, and franchises to the City of Denver for $9 million.[28]

The city rejected the price, which was, after all, nearly half what the United States would pay Spain for the Philippines the same year. The city did offer to buy some individual parts of the company's properties with the intention of incorporating those parts into a unified system of its own. But David Moffat and Walter Cheesman, as leaders of Denver Union, maintained that they would "sell nothing less than the entire system," sticking with the initial $9 million offer. "Must Have All or Nothing," ran the headline in the *Rocky Mountain News*, even as the story quoted a more hopeful, if also more cynical, analysis from Joel W. Shackelford, president of the Board of Public Works: "The company is a good bluffer. . . . People are not apt to name their lowest prices at the start. The water company had always been a hard bluffer and we must make some allowances for that." Whether or not Moffat and Cheesman were bluffing (and they were not), with their combined experience in managing large railroad corporations, as well as their records of achievement in the banking industry, they were formidable competitors and rivals. The city was in for a tough negotiation.[29]

As the jockeying continued, the city's engineer assessed Denver Union's property and water rights and appraised their value at more than $3.75 million. In November 1899, the city council asked citizens to vote on a single $4.7 million bond that could then be dedicated to either the purchase of Denver Union's property or the construction of a separate municipal water system. A notably low turnout of taxpaying voters approved the bond on November 29, 1899.*

But in 1901, the US Circuit Court declared the bond election invalid on technical grounds because, among other issues, it was unlawful to combine in one initiative the two possible courses of action, which should have been submitted separately. The court's decision kept the city from taking action on the water issue until another approach to securing the funding could come into play. In 1904, however, the city gained a powerful tool for building a municipal water system when voters approved home rule for Denver in Article XX of the state constitution, granting the city council the authority to supersede state laws on local matters. In practice, the Home Rule Amendment (or the Rush Amendment, both of which were more popular names for it than Article XX) gave the city the power to take resources both within and outside of its borders, providing it with the necessary legal mechanism to build a water system far beyond its city limits.[30]

As Denver Union's twenty-year franchise approached the end of its term in 1910, the company and the city resumed negotiations. Following the terms spelled out in the 1890 agreement, an impartial group of appraisers, composed of five renowned engineers, were hired in 1907 to assess the value of Denver Union's assets. In March 1909, the engineers concluded that the water company was worth $14.4 million; the addition of Cheesman Dam had significantly increased the value.[31]

---

*Jerome Smiley asserts that, despite the large dollar figure, the low voter turnout on the bond election indicated that "the people were rather singularly indifferent" about the water question at this time. Although the roles of eligible voters were limited to only taxpayers and the election was a special one not held in conjunction with a general election, Smiley's comment raises an interesting question about whether the public demand for municipal water was as popular as some suppose; the city council and other city leaders (through personal self-interest or desire for public benefit) may have driven the demand for municipal water. Still, among those who did vote, a solid majority of 5,420 to 2,976 carried the outcome. See Smiley, *History of Denver*, 935-36.

Both the city and the water company balked at that number. City engineer John B. Hunter asserted that the city could replicate Denver Union's water plant for $8 million, while Denver Union's directors and officers thought the estimate of the value of their system was far too low.[32]

Despite the unsettled issue of purchase price, the Progressive desire for a municipally owned water supply gained ground. After suffering a setback in the creation of a business-friendly 1904 Denver City Charter, Progressive reformers regained political clout when they succeeded in enhancing the regulation of municipal railroad and gas service in 1906, hoping that this would set a precedent for the expansion of the city's role in water supply.[33]

A number of Progressive reform organizations—including the Ward Water Users' Association, the Labor Water League, the Christian Citizenship Union, and the Municipal Ownership League—came together as the Citizens' Party to make the case for municipal ownership.[34]

The Citizens' Party and the *Rocky Mountain News*, championing the Progressive cause, urged local citizens not to renew the water company's franchise, and the Citizens' Party put forward a ballot initiative with that aim. The effort paid off in the May 1910 election, when taxpayers voted against giving the company a new franchise and instead voted in favor of creating the Denver Public Utilities Commission, with the power to acquire and manage a water system for the city. The successful amendment also required the commission to investigate yet again the prospect of buying Denver Union's system and authorized them to offer $7 million for the company's plant, spelling out that this offer had to be accepted before July 1, 1910. If the company rejected the city's offer, the amendment mandated a special election to approve the allocation of $8 million in bonds to build a separate municipal water system.[35]

In control of the company since Cheesman's death in 1907, Denver Union president David Moffat and the company's stockholders rejected the city's offer, citing the previous appraisal of their property at more than $14 million. The following September, voters approved the $8 million bond to build the city's own system.[36]

But litigation pursued by Denver Union's bondholders stalled further action. A series of court battles led all the way to the US Supreme Court, and in 1913 the Court ruled against the company. Amid all the contentious litigation, Denver Union went on serving the city in spite of the fact that its old franchise had expired and no new franchise had replaced it.[37]

In August 1913, with the city empowered to proceed, thanks to the Supreme Court decision, the Denver Public Utilities Commission appointed a three-man board of engineers to study the feasibility of creating the city's own water system. The committee was assigned to survey the present value of the Denver Union Water Company's holdings, the projects proposed for furnishing water from the Western Slope to Denver, and the various storage projects on the South Platte River and its tributaries. After this assessment, the engineers were to estimate the cost of acquiring or constructing a complete water system for the city of Denver by any of the methods proposed. Edwin Van Cise, the president of the Denver Public Utilities Commission, was by this point communicating with the company in uncompromising terms: "We are not," he wrote, "acting as private citizens with sympathy for your stockholders, but as public servants with duties to perform."[38]

The Board of Engineers delivered its report to the Denver Public Utilities Commission on January 16, 1914, appraising the net value of all the Denver Union Company's property and assets at just over $10 million. The report also catalogued the numerous challenges involved in constructing a new water system. With those challenges in mind, the engineers argued against having Denver build its own water system while also recommending that the city consider purchasing its own Western Slope water rights, identifying the Fraser, the Williams Fork, and the Blue Rivers as promising sources of supply.*

As an example of the remarkable variation at work in this world of estimating values and costs, in December 1916 the engineering firm Van Sant-Houghton delivered its conclusion that constructing a municipal water system from scratch would cost the city at least $27.4 million.[39]

A seemingly inauspicious turn of events, apparently escalating the conflict, turned out to set in motion a chain of actions leading

* In August 1914, the utilities commission had also taken a small step toward securing its own water supply by contracting with Antero and Lost Park Reservoir Company to purchase the Antero Reservoir and the High Line Canal. Due to litigation over issues of title and ownership, however, the sale was temporarily enjoined, and the Board of Water Commissioners was not actually able to take over control of Antero Reservoir and the High Line Canal until 1924. See "Antero Contract with the City of Denver," August 21, 1915. On transmountain diversions, see James Lee Cox, "The Development and Administration of Water Supply Programs in the Denver Metropolitan Area" (PhD dissertation, University of Colorado–Boulder, Department of Political Science, 1965), 244–45.

to the resolution of the whole wrestling match. In March 1914, the city council passed an ordinance to "regulate and fix" Denver Union's rates, imposing a 20 percent reduction on the company's rates. The company responded by suing the city, rejecting the city's reduction of its income as unjust. Colorado Springs attorney W. J. Chinn was appointed as a special master in the litigation over the legitimacy of the company's rates. In October 1915, Chinn concluded that the city's prescribed rates did "not permit the Water Company to earn for the use of its property such compensation as is fair to it and the city." Considerably more consequentially, as part of his investigation, Chinn also asserted that "$13,415,899 is the fair and reasonable value of the property [owned] by the Water Company." Chinn's evaluation was based in part on an earlier US Supreme Court case that allowed companies to include "going concern value" (meaning the value that accrues to an ongoing solvent business venture) into the determination of property value.[40]

Although the original lawsuit began in a dispute over water rates, Special Master Chinn's appraisal of the company's value was his most significant finding. Just six months after the ruling, the city entered into a tentative agreement with Denver Union to purchase the entire water system for the court's appraised price. However, the completion of the purchase would await one more ruling from the US Supreme Court. On March 4, 1918, the Court affirmed Chinn's findings and thus cleared the way for the formal purchase on November 1, 1918.[41]

⧫⧫⧫

In 1909, Denver city officials had scoffed at the water company's $14 million price tag. But six years later, the city council agreed to pay more than $13.4 million without much discussion. The Van Sant-Houghton engineering report, with its estimated $27 million price tag for an entirely new water system, probably helped persuade the council members to reconsider their positions. In a bond election on August 6, 1918, Denver citizens approved the purchase of the Denver Union Water Company. After decades of disputes and negotiation, the city finally had its own municipal water supply system.[42]

Approving the bonds to purchase the water company, taxpayers also voted in favor of a charter amendment that created the five-member Board of Water Commissioners who would oversee and control the new municipal system. Water commissioners were nominated by the mayor and confirmed by a public vote. Bylaws mandated that board members served staggered six-year terms. In the early years, the commissioners received annual renewal of their appointments with no term restrictions. From the beginning, the goal was to structure the Board of Water Commissioners to limit political influences. Although the mayor appoints commissioners, neither the mayor nor the city council has the power to remove them. The staggering of terms, moreover, makes it difficult for one mayor to appoint a majority.[43]

As the conflicts and contests of the early twentieth century settled down, the Denver Water Board—a term that came to refer both to the Board of Water Commissioners and the agency they directed, although that agency is also commonly referred to as the Denver Water Department and, simply, Denver Water*—emerged as a new entity with the responsibility for setting water rates and purchasing, operating, and maintaining water facilities, as well as securing water rights for use by the city. Since its inception, the Water Board has, by design of the Denver City Charter, operated as a nonprofit governmental organization, turning over all revenue to the treasurer of the City and County of Denver. The treasurer then deposits that revenue into the Water Works Fund, which is accessible only to the water commissioners and audited annually by the Denver city auditor. Any hydroelectric power generated at Denver Water Board facilities must be sold at wholesale rates. Perhaps the key element of its charter is this: while the Water Board has the authority to provide water to communities outside of Denver, Denver Water's first responsibility is always "to provide an adequate supply of water to the people of Denver." In practice, outside sales were sometimes necessary in order to develop coalitions of communities large enough to marshal the resources needed for big projects. But the primary mission would always be clear: to provide "adequate" water to the capital city.[44]

---

* A note on terminology: The city water agency was originally called the Denver Municipal Water Works, although that name did not last long. A variety of monikers for the agency have, however, had a great deal of staying power, subsequently including the Denver Water Board, the Denver Water Department, and Denver Water (which is the most common iteration at present). Although the terms carry slightly different connotations related to the era in which they were prominent, this manuscript uses them synonymously when referring to the agency and its commissioners.

Between 1894 and 1918, the Denver Union Water Company built the foundation on which the Denver Water Department came to rest. Led by Cheesman and Moffat, Denver Union completed the world's tallest dam in 1905 and also expanded its capacity to deliver abundant water to the city's growing neighborhoods. In 1907 alone, for instance, the company built a new pumping plant in present-day Wheat Ridge and constructed another covered, concrete-lined downtown storage basin beside the existing Capitol Hill Reservoir near Congress Park.[45]

Over the years of its existence, the company also worked to improve the quality of its water by building filtration galleries, covering conduits to convey water from storage reservoirs, and chlorinating water at the Marston Treatment Plant.[46]

While great improvements in the city's water system took place in this era, attention and energy also went into long-running political and legal battles over who should provide water to the citizens and at what cost. In 1918, as the Denver Union Water Company became the publicly owned Denver Municipal Water Works, the era of private water companies in the city came to a close. The new institution had little opportunity to pause to rest after its creation. With a city population approaching 257,000, the Denver Municipal Water Works was under considerable pressure to expand the water system, to maintain and prepare aging units within the system, and to reach beyond the Continental Divide.[47]

While the operators of the water system had focused on anticipating and preparing for scarcity by acquiring and storing water, in the next two decades, flooding would also pose formidable challenges. A major flood along the South Platte River inundated Denver in June 1921. Although only two people lost their lives, so many roads and railroad tracks were submerged and so many bridges washed out through the region that many factories were forced to shut down until supply lines could be reopened. The Water Board's infrastructure sustained such heavy damage—requiring approximately $200,000 in repairs over the next few months—that the city faced water shortages. Board president Finlay MacFarland urged citizens to cut back on their use of water until the system could be patched, though he stopped short of imposing restrictions on the irrigation of gardens and lawns.[48]

Flood management was (and is) outside Denver Water's mission and legal responsibility. Indeed, controlling floodwaters could stand at cross-purposes with the objective of filling reservoirs close to capacity to ensure the ability to deliver water throughout the dry season. But the impacts of floods remained damaging and expensive. While the contest to designate the institution that would serve as the city's water provider may have been resolved with the Denver Water Board as ultimate victor, the difficulty of securing an abundant and reliable water supply in a land with a fluctuating climate was a chronic condition with which Denverites would continue to grapple. By 1920, the city's population had reached 256,401, and the new Denver Water Board was providing water to 270,000 customers (households and businesses combined).[49]

In western cities, the historian Carl Abbott writes, "water flows through kitchen faucets, fire hydrants, and factory valves because feats of engineering are matched with creative institutional arrangements."[50]

In Denver's case, in the era around 1900, the designing and establishing of "creative institutional arrangements" involved a considerably more tangled and prolonged process than the "feats of engineering." The lesson of that contrast was not subtle: in designing and building dams, reservoirs, and distribution systems, engineers took on a very strenuous enterprise, but the project of coordinating voters, public officials, and corporate owners presented twists, turns, complexities, and reversals that made it look as if the engineers had drawn by far the easier task. By several acts of fortune, Denver's early efforts at securing water from a distant location had gone much more rapidly and smoothly than comparable efforts by San Francisco and Los Angeles. But the struggle to create the Denver Water Department and Board of Water Commissioners gave fair warning: in the twentieth century, Denver stood a very good chance of catching up to its California counterparts in the accumulation of experiences of contention, litigation, controversy, and difficulties in dealing with federal authority. Building Cheesman Dam involved no significant litigation; building the Denver Water Department involved litigation aplenty, including two cases decided by the US Supreme Court.

In the struggles over municipal ownership, the majority of the voting citizens of Denver had become functioning antimonopolists, supporting governmental power over the control of water by a private company. In hindsight, a question unavoidably comes to mind: if the discontented citizens of Denver did not like the centralized power of a company, were they likely to find contentment with the centralized power of a municipal agency? Or if the residents of the City and County of Denver came to feel that this new centralized power served their interests, how would the residents of other cities and counties in the State of Colorado interpret and respond to that institutionally mandated favoritism? When the interests of Denver collided with the interests of other communities, it seemed certain that the concentrated authority residing in the Denver Water Board would trigger resistance and anger in other communities. The shift to municipal ownership had changed the terms of concern over centralized power, but it had by no means eliminated the debate over its fairness and equity.

During the first decades of the twentieth century, from the construction of Cheesman Reservoir to the creation of the Water Board, a profound and consequential shift had occurred; the citizens of Denver had come to expect that the Water Board would always provide them with an abundant and safe supply of water. A historical era in which people dug their own wells or hauled their own water in buckets faded from memory. Denver residents came to think of the miracle of turning a handle in the kitchen to produce a flow of water as a perfectly normal feature of life.

As they tended their yards, shopped for fresh produce, waited for their weekly visit from the sprinkling wagon, and enjoyed the countless other uses of water that had quickly been reclassified as unquestionable needs, westerners surrendered to amnesia. Knowledge of what it was like to live under the full powers of aridity became the exclusive and fading memory of the first generation of Denverites. Later arrivals, and some of the old-timers, looked at a radically changed, much greener landscape, made possible by the transportation of water from one site to another, and classified what they saw as normal and expected. The manipulation of water made Denver happen in terrain that had seemed incapable of supporting a concentrated human population of any size. Occasional dry spells might rattle confidence, reminding Denverites that they remained vulnerable to drought. But more and more, locals took to considering periods of low precipitation and reduced stream flow as exceptional and atypical and to thinking of a ready supply of water as a dimension of life they had the right to expect. The Era of Improbable Comfort Made Possible by a Taken-for-Granted but Truly Astonishing Infrastructure was taking shape.

Part Two

Colorado and the Consequential Rectangle

The men who created this state
Gave an ungainly shape to our fate.
It hardly seems fair—
That the love of a square
Was their strongest and most forceful trait.

Chapter Three

Water Development: "The Plot Thickens"*

Many Denver antagonists regard Glenn Saunders as a Machiavellian plotter. It is a fact that he is one of Colorado's shrewdest water lawyers. . . . He is tough, devious, and dedicated. . . . Saunders has a way of making people mad at him and there are times when his public utterances have been viewed as indiscreet. . . .

Close associates of Saunders say he isn't the ogre his critics depict. He simply doesn't have time to be a goodwill ambassador while battling for Denver's interests, and he firmly believes that Denver has always acted in good faith and abided by all laws.

—Bert Hanna in *The Denver Post*[1]

The foothills area from Colorado Springs to Fort Collins is one of the perfect climates for thinking and living in the world. It is constantly attracting thousands of people who must have water.

—Glenn Saunders[2]

When it comes to the history of western water, there is a select and short list of celebrities. Two of the best known are John Wesley Powell, explorer, naturalist, and author of the *Report on the Lands of the Arid Region of the United States* in 1878, and Floyd Dominy, legendary commissioner of the Bureau of Reclamation in the mid-twentieth century. While they were unquestionably historically significant figures, they both owed some portion of their celebrity status to their good fortune at falling into the hands of gifted and persuasive writers. In *Beyond the Hundredth Meridian*, Wallace Stegner told readers the tale of John Wesley Powell, a heroic and farsighted figure of the nineteenth century who recognized the limits of the West's water and courageously defied the bullies and blowhards who insisted that the region's resources were infinite,

* Title of speech given on February 2, 1939, by Denver Water special counsel Glenn Saunders.

and Stegner's readers have, for decades, applauded and cheered Powell. In *Cadillac Desert*, Marc Reisner told his readers the tale of Floyd Dominy, a twentieth-century reincarnation of those bullies and blowhards whom Powell had been unable to defeat, and Reisner's readers have, for decades, booed and hissed Dominy.

The name Glenn Saunders should join the select and short list of water history celebrities—because his work transformed the western landscape, and because his complexity puzzles us and makes us think. Neither Stegner nor Reisner could have handled him in a manner that would have given their readers a clear cue as to whether they should boo and hiss, applaud and cheer, or offer some jumbled combination of those two lively forms of expression.

The year 1918 carries double weight in the history of Denver Water: first, in the autumn of that year, the City and County of Denver purchased the Denver Union Water Company, and the Denver Water Board came into being; and second, a few months earlier, a thirteen-year-old Glenn Saunders got his first job through "a near neighbor" who happened to be the chief engineer of the Denver Union Water Company. Few summer jobs held by teenagers have launched an institutional tie of greater consequence for a person and for an entire region. "In the summer of 1918," Glenn Saunders reminisced seven decades later, "I was employed to watch the float gauges on the clear water basins in the Capitol Hill Pump Station in Denver, which supplied water to everything east of the South Platte River." The float gauges had to be watched carefully because "there were many wood-stave conduits" in the system. "If one of them broke," Saunders said, "it needed to be known immediately," and thus it was the job of the gauge-watcher to send out an instant alarm when a noticeable drop in pressure indicated trouble.[3]

After he completed law school at the University of Michigan in 1929, Saunders returned to Colorado and worked, first, in the city attorney's office with a special assignment to the Denver Water Board. In 1937, "he took over full responsibility for the board's water rights development."[4]

Over the next decades, he watched over and tended to the whole system as vigilantly and intensely as he had once watched over and tended those pressure gauges, sending out an instant alarm when he detected signs that the system was in jeopardy. He was an unflagging advocate for the cause of securing water for the clients of Denver Water. The well-being of Colorado rested on the

Glenn Saunders stands beneath a bridge near Kremmling in November 1936, looking out at the Colorado River.

well-being of its capital city, Saunders believed, and thus Denver's access to Western Slope water dwarfed the concerns of farmers and nature lovers. His claims for Denver's importance went nationwide in scale: "Our growth," he said, "is as much to the advantage of the United States as it is to Denver."[5]

Saunders played a crucial role in the city's major transmountain diversion projects from the Fraser, Williams Fork, and Blue Rivers in the Colorado River Basin. His single-minded energy was so striking and so effective that he was at once a force reshaping nature and a force *of* nature. Whether he battled his counterparts on the Western Slope who wanted to keep water in its basin of origin to support their own plans for economic development, or faced off against environmental advocates who wanted to put the brakes on development so that water would flow unrestricted in streams and rivers, Saunders used every legal, political, and rhetorical weapon he could devise. When it comes to a wholehearted and unyielding campaign of a public official on behalf of the water-managing organization that employed him, Saunders played at the top of the major leagues. He was one of the most forceful and, accordingly, one of the most controversial figures in western water history.

And, in a proposition that some might initially dispute, he was a complicated person, impossible to categorize in a simple way.

For residents on the Western Slope, however, placing Glenn Saunders in a simple category presented itself as one of life's easier tasks. Especially in the 1950s and the early 1960s, the language used by his opponents to characterize Saunders calls into question the premise that Americans once operated in a serene and civil arena of public discussion, a treasured tenet of twenty-first-century nostalgia. Saunders, wrote William Nelson in the Western Slope's *Grand Junction Sentinel* in 1955, refused to "accept any limitations." Under his direction, "Denver has attempted every means that came to mind in the past thirty years to steal Western Colorado water." Denouncing Saunders's "usual ruthless tactics," the *Grand Junction Sentinel* editors said that "his screaming is infantile and silly," a form of posturing intended "to make Denver citizens think he is a gallant knight riding a white charger to defeat those selfish barbarians west of the mountains." There would be no chance of statewide unity, the *Sentinel* said, "as long as Saunders runs the Denver Water show on the theory that might makes right and to the devil with anyone who disagrees."[6]

Of course, when it came to harsh and sharp-edged remarks, Saunders gave as good as he got. Headed toward one showdown, when Saunders insisted that the origin of the conflict lay in the "selfish motives" of the Western Slope advocates and their "narrow and provincial" point of view, his use of the adjectives *selfish, narrow,* and *provincial* demonstrated the enthusiasm that pots often have for calling kettles black. Both Saunders and the Western Slope folks put their home areas first; the Western Slope people just had the poor luck to be defending the wrong subregion. Even when he spoke of legal transactions that left experienced lawyers befuddled, he used terms of seeming simplicity: "Western Colorado interests" conducted "a continuing effort" to "prevent the development" of "water to which we are clearly entitled." It is interesting to contrast the certainty of Saunders's assertion with a description of this particular showdown (the litigation leading up to the Blue River Decree of 1964) by an astute and thoughtful reporter: "Many distinguished lawyers and water experts are confused by all the ramifications of the Blue River water battle." The more confused they were, the more Glenn Saunders scented opportunity.[7]

Some episodes of Saundersian bluntness leave the reader nearly breathless. On one occasion, Saunders was taking a deposition from the executive director of Trout Unlimited, Robert Weaver. To Saunders, the fact that Weaver had not been born in Colorado meant that he had little standing to criticize the practices of the Denver Water Board (a curious position in itself for a man who spent his career supporting the need for more water in order to accommodate the needs and desires of newcomers to the state). This line of questioning brought Weaver's attorney, J. Kent Miller, into a memorable exchange with Saunders. Trying to place Saunders's belief in the legitimacy of the native-born into a longer historical context, Miller asked him, "Did the Indians require a permit from your ancestors to settle in Colorado?"

This query unleashed a legendary episode of outspokenness.

"No," replied Saunders. "We killed them off and conquered them because we were stronger and more aggressive and if they rise again, I am willing to do that again."

Miller may have felt a moment's tilt toward speechlessness, but he pressed on: "So your property rights are superior because of your ability to conquer?"

His ability to conquer, Saunders explained, derived from "my superior intellect and physique, this is exactly right, the same as your folks. That is why we are here, too. We are superior people and we should be proud of it and use it well, and I am not ashamed of being superior to some of the people I have seen in this world."[8]

This exchange, it may be useful to point out, occurred in *1974*, not 1874. In fact, Saunders was a fluent speaker of the language of nineteenth-century westward expansion.

"As we get more civilized," he said in 1944, "we'll need more water. It always happens that way." Even when twentieth-century words like *ecology* entered his expressions, they produced no noticeable softening of his point of view. "We are living . . . in an uninhabitable area," Saunders said in 1971. "We changed the ecology to get what we have here, to improve our environment."[9]

Readers may now be puzzling over my earlier assertion of Saunders's complexity. Saunders may seem to fit every specification as an uncomplicated advocate for the building of dams, reservoirs, tunnels, and treatment plants for the benefit of the Denver Water Department and its service area. At this point, he may seem to be the embodiment of W. E. B. DuBois's great aphorism, "It is as though Nature must needs make men narrow in order to give them force."

Which makes this a good time to turn to the complications.

In a textbook of environmental history published in 1999, historian John Opie quotes a person he only identifies as "a local Colorado attorney" named Glenn Saunders. Opie is a member of an earlier generation of environmental historians who were themselves committed supporters of environmentalism and whose historical writings thus made a clear distinction between bad guys (heartless pursuers of profit who forced their will on nature) and good guys (principled souls who opposed the despoliation of nature). Thus, when he quotes at length from this fellow named Saunders, Opie's approval of the sentiments in the quotation is unmistakable. In this passage, Saunders was expressing his opinion of a new dam, the Narrows, proposed by the Bureau of Reclamation to benefit a comparatively small number of farmers on the eastern plains of Colorado. (Readers in the habit of skipping over long block quotations should make an exception for this one!)

> The people who support these boondoggle projects are always talking about the vision and principles that made this country

> great. "Our forefathers would have built these projects!" they say. "They had vision!" That's pure nonsense. It wasn't the vision and principles of our forefathers that made this country great. It was the huge unused bonanza they found here. One wave of immigrants after another could occupy new land. There was topsoil, water—there was gold, silver, and iron ore lying right on top of the earth. We picked our way through a ripe orchard and made it bare. We've been so busy spending money and reaping the fruits that we're blind to the fact there are no more fruits. By trying to make things better [by building dams], we're making them worse and worse.[10]

In its wholehearted, no-holds-barred condemnation of American exploitation of nature, this was a quotation certain to earn the approval of historian and environmental advocate John Opie. Asking a few background questions about the speaker would only have led to trouble, ruining a perfect quotation.

In multiple field tests, I have asked audiences to listen to this quotation and, from it, deduce where Saunders registered in the spectrum of opinion on nature and its uses. This exercise will consistently yield the answer, "He was a strong environmentalist and opponent of dam building." When told that Saunders actually spent his career getting dams built and fighting environmentalists, these audiences are presented with one of life's prime opportunities to reckon with the complexity of historical figures.

The passage that charmed John Opie most likely came to his attention through a reading of the very popular book on water projects in the West, Marc Reisner's *Cadillac Desert.* Unlike Opie, Reisner did pay attention to the bigger picture of Saunders's identity, recognizing him as "the man who championed water development for fifty years in Colorado." To explain the curious relationship between the goals that drove Saunders's career and the very lively quotation that John Opie seized upon, Reisner offers the proposition that the Narrows project was so terrible, particularly in its cost–benefit ratio, that it made a new man of Saunders. "As he readily admits," Reisner wrote, "it changed his whole way of looking at things."[11]

It would be nice to know if that were true. Before we accept it, it would be important to remind ourselves that the Narrows Dam was a competing project that might intrude on Denver Water's plans. Furthermore, Saunders had always been intensely critical of the federal government, and he and the Bureau of Reclamation struggled for years over the operation of the bureau's Green Mountain Reservoir on the Blue River. So, while Reisner's change-of-attitude explanation is thought provoking, a continuity-of-attitude hypothesis may be more convincing, since giving the Bureau of Reclamation fits had been a long-running enthusiasm and a source of considerable satisfaction for Saunders. Similarly, Saunders held a clear preference for urban and domestic use of water over agricultural irrigation; there was thus nothing particularly novel in his challenging a dam that would offer no advantage to Denver.

When it comes to peering into the human soul, the equivalent to X-rays, MRIs, or CT scans awaits a higher level of engineering. The project of figuring out what lay behind Saunders's late-in-life condemnation of the American exhaustion of natural resources will be an endless one. It is unmistakably true that Saunders had an expert theatricality (not uncommon in gifted attorneys) that allowed him to shift rhetorical style and tone as some opportunities closed down and others opened up. Was his spirited reappraisal of the pioneer heritage of the United States a matter of Saunders choosing a new script for the mid- and late 1970s? Or was he simply enjoying the opportunity to pick up a two-by-four and whale away at his old rival, the Bureau of Reclamation?

Whatever Saunders was doing, he was not cooperating with any simple efforts to categorize him.

In 1965, eleven years before the Narrows Dam controversy, Glenn Saunders put on public record the most striking evidence of his complexity. The setting was, at first glance, entirely improbable: Saunders was in Vail, giving a principal speech at the second annual Colorado Open Space conference, sponsored by the Colorado Mountain Club. The invitation itself deserves a moment of wonder: the most visible advocate of damming and diverting Western Slope rivers in remote, scenic locales was going to give a keynote speech to a group of environmentalists fighting to preserve the open spaces in which Saunders and his allies wanted to place dams.

The wonder only deepens when we contemplate the content of Saunders's speech. He recognized the devotion of his audience to the preservation of "natural beauty" and to "conservation." With equal directness, he acknowledged "his lifelong devotion to developing the water system for Denver at the expense of the most beautiful streams in Colorado."

Halfway through the newspaper's report of the summary of this speech, nearly half a century after Saunders delivered it, one moves to the edge of one's seat, wondering what on earth will happen next. In a room seeming to contain people so dramatically in opposition, would Saunders be shown to the door while the moderator of the conference announced an unscheduled coffee break, during which a speaker who made a better fit to the occasion would undergo an emergency recruitment?

On the contrary, Saunders and his audience stuck it out. He said that the attendees at the open space conference were the kind of group "that can give us a better world." In solidarity with them, he declared his belief that "the world is a temple left to us as a sacred trust." But the sacred trust could not be honored until human beings recognized that they themselves were the greatest threat to the earthly temple. "As long as we have so little intelligence as to limiting the population," Saunders declared, "we cannot expect the kind of society we want." Americans believed that there must be "more people, more prosperity, more and more and more and more. . . . Never is there the civilized idea," he lectured, "that at some point we should be able to stabilize population, economy, numbers—without destroying the attunement of the human race to the whole universe." If Americans truly wanted to create the kind of society they desired and protect the environment, Saunders explained, the solution was as clear as the alternative: "Reach for the cause bringing about the horrible condition. Otherwise people like me will be out to destroy the natural beauty of the country."[12]

Consider the context of late September 1965. Saunders gave his speech to the Colorado Open Space conference on September 25, and the newspaper story summarizing it appeared on September 26. Three days later, another story appeared, with the headline "Leave Granted to Saunders by Water Board." It would not be the wildest guess to speculate that learning of Saunders's speech at Vail had annoyed the members of the Denver Water Board to the point where they thought that they might enjoy a little break from his company. Until people could limit their numbers, he had said to the open space advocates, "grim necessity requires me to destroy the basic values for which you all strive." It was and is hard to take in the idea that the most widely recognized official of the Denver Water Board had said such a thing on public record. It might also be reasonable to suppose that the board was not thrilled that their staff attorney and de facto spokesman also had a private practice beyond the commissioners' control. And yet the stated reason for granting a leave of absence was entirely unrelated to the speech or Saunders's practice. Saunders was a "chronic asthma sufferer," and the leave was given for "health reasons" because "board members felt he was near collapse." This timing is striking. When Saunders spoke at Vail, did his blunt honesty there arise in part from a sense of his own frailty and mortality? Can such a question ever receive even a hint of an answer?[13]

Readers now finding themselves moved to rescue Saunders from historical obscurity by celebrating him as a farsighted and courageous prophet of the dilemmas of an expanding human population should, however, reckon with the next layer in the man's complexity. In an article in *The Denver Post* in 1967, reporter Lee Olson wrote of Saunders's thoughts about the burdens posed by human numbers: "He concedes there is a population problem in the long run. He sees birth control, coupled with deliberate efforts to improve the population, as logical." Saunders told the reporter that "the day is coming when we will control not only the volume but the quality of our population."[14]

Saunders seems to have been an enthusiast for eugenics, the deliberate breeding of a human population in favor of certain traits and characteristics. This is a stance that once had many admired supporters, with Theodore Roosevelt as one of the most prominent, voicing his concerns about the "race suicide" of white Americans who were failing to reproduce with proper competitive pep. By the 1960s, the casting of some groups of people as superior and some as inferior, in the manner of Saunders's lively remarks on the conquest of Indian people, had become a matter of considerable discomfort and dismay. Curiously, like Roosevelt, who had struggled as a child against illness and frailty, Saunders claimed status as a specimen of genetic superiority while enduring a lifelong struggle with frailty. But he was also, in the Rooseveltian manner, an athlete, dominating the action on Denver handball courts as he often dominated the action in courts of law.

His respiratory problems did not interfere with living a long and productive life, and Saunders continued to work long after his retirement. In writing the history of a public official, stating when they took their offices and when they left those offices is usually a matter of simple and elemental factuality. Not so, however, with the complicated Glenn Saunders. The process of his departure from the position of lead counsel for Denver Water is hard to reconstruct with precision. After his 1965 leave of absence, in February 1966, *The Denver Post* carried the headline "Water Counsel Eased Aside." *Eased aside,* of course, is quite a different verb from *fired* or *terminated* or *dismissed.* "The Denver Water Board graciously eased its veteran attorney, Glenn G. Saunders," the reporter summarized this transaction, "out of any policy-making or managerial functions and made him an outside legal consultant with offices removed from the water department." He would, however, "enjoy the same salary." Saunders said "he was happy with the board action because it will help him to regain his health." The board's goal was to narrow the scope of his authority: "He will continue to represent the department," the *Post* said, "but he won't have a voice in how the department is operated," a doubtful prediction given Saunders's forceful personality and deep interest in the affairs of the Denver Water Board. It was an overly hopeful notion and perhaps an overly rosy rendition of the truth. Years later, an obituary for Saunders recounted the story rather differently: "In 1966, Saunders sent the water board into turmoil when he quit and took with him the city's best water attorneys to form the 17th Street law firm of Saunders, Snyder, Ross & Dickson." After formal declarations of an "easing aside," "that firm sold its advice back to the water board for hundreds of thousands of dollars a year."[15]

Whatever the true circumstances of his departure from his post, Saunders's Western Slope opponents were not comforted or reassured by news of Saunders's shift in status. In an editorial, the *Grand Junction Sentinel* repeated its usual characterization: "Saunders was long considered to be a virtual dictator in water board affairs." A technical change in his status meant little: "Outspoken, shrewd, and ruthless," Saunders would "never relent in his efforts to get as much water as possible for the capital city." Rather than being fooled by an administrative shuffling, "western Colorado organizations must not relax their vigilance one minute."[16]

In truth, it was not easy to detect much in the way of change in Saunders's visibility or in his central role as Denver Water's legal representative. Three years later, the board formally recognized Saunders's retirement at age sixty-five, which would certainly have been a peculiar ritual to observe if Saunders had actually severed his connection to the board in 1966. "He's unlikely to head for a rocking chair with his drive and dedication," the board said, prophesying accurately this time, in its formal recognition in 1969; "it's likely he will remain an important force in Denver's water future."[17]

Even though *The Denver Post* carried a headline announcing "Water Attorney Will Retire" on August 27, 1969, there were many occasions in the years *after* 1969 when Saunders represented the Denver Water Board in court, presented a plan for a particular project to the board, or announced the department's policies. In 1973, four years after Saunders's ostensible retirement, when several environmental groups and cities came together to file a suit challenging the legitimacy of Denver Water's transmountain diversions, Saunders asserted the city's rights and dismissed the arguments of the suit. "The Board has the authority to divert . . . water from the Western Slope," Saunders was quoted in the article, with no evident slippage either in his authority to speak for Denver Water or in the force with which he spoke. "He added that Denver has no obligation to consult Western Slope voters on its water rights decisions." When this suit came to trial two years later, the *Rocky Mountain News* reported that the "young attorney" J. Kent Miller was "pitted against the savvy Glenn Saunders, 71," who had been the Water Board's "special counsel for forty-five years." Nine years after he had been "eased aside," and six years after his official retirement, Saunders's level of activity on behalf of his employer did not seem dramatically diminished.[18]

Glenn Saunders was often singled out as the public face of the Denver Water Board, and his outspokenness made him a very visible figure. "Glenn Saunders IS the Denver Water Board," the editors of the *Grand Junction Sentinel* wrote, encapsulating a widely shared opinion. "Whatever he decides, the Board does." This was an exaggeration; Saunders was only one person—albeit one very active, visible, and audible person—working for a big and complicated organization. But his historical role in the expansion of the water supply and delivery system for the city of Denver was genuinely and provably enormous. Thousands of people who have lived, worked, and prospered in Denver are indebted to a man whose name they do not recognize. In truth, the degree of his departure from public memory is unsettling. One thorough, detailed, and reliable history of Denver, for instance, published in 1990, the year of his death, does not have the entry *Saunders, Glenn* in the index.[19]

At the time of his "retirement," the Denver Water Board noted that the existence of "a metropolis" in "a semiarid location" with

Glenn Saunders at age fifty-nine in 1963. For many Coloradans of the era, his was the face of the Denver Water Board.

"meager water supplies" was "a tribute to men such as Glenn G. Saunders." "Friends and enemies alike," the board said, "respect Saunders and most would label him one of the nation's most brilliant water attorneys." In a resolution passed September 15, 1969, the manager of Denver Water and the board declared that Saunders's "zeal and unlimited dedication to the best interest of the Water Department warrant[ed] the establishment of an appropriate memorial." Thus, the board resolved, "The Board of Water Commissioners of the City and County of Denver shall, and does hereby, authorize the selection and designation of an appropriate water facility in the Water Department to be known and named in honor of GLENN G. SAUNDERS." The next step would be the "selection of an appropriate water facility" to be "named in his honor."[20]

Other priorities seized the attention of the Denver Water Board, and the resolution to select a facility to name after Saunders fell by the wayside. Indeed, it is difficult to conjure up a statement that would fit in the constrained space of a brass plaque or a stone panel and, at the same time, adequately address Saunders's complexity. Moreover, by the 1970s, the rapid pace of change in American attitudes toward nature would have made it a challenge to find the right tone and emphasis with which to celebrate the heritage left by a man who had offered for the public record many very blunt condemnations of popular trends of his times. In 1958, contemplating early moves toward the Wilderness Act of 1964, Saunders asked, "How can any right-thinking American give serious consideration to wilderness legislation which will hamper [necessary] water development?" Proposals for wilderness, he proclaimed, offered "but one striking example of the damage which will be done to our economy, if we fail to curb the present threatening trend to restrict use of our federally owned natural resources for the exclusive benefit of our people." As the environmental regulations of the 1960s and '70s announced Americans' changing views of the environment, Saunders did not fall into step. In 1975, speaking of the impact of the National Environmental Policy Act of 1970, he reiterated his view that "requiring all these unnecessary environmental impact statements is upsetting the entire economy of our country." [21]

With his name attached to statements such as these, the scripting and orchestrating of the dedication ceremony for a Saunders Tunnel, a Saunders Reservoir, or a Saunders Treatment Plant would have strained the talents and shaken the morale of the most fluent, nimble, and conciliatory of speechwriters and event planners. And yet those speechwriters would also have been able to draw on a stockpile of quotations from Saunders in which he spoke for collaboration and cooperation. "All efforts should be made to try to resolve the controversial issues," he said in December 1963. "Avoidance of a long and bitter court battle would be in the best interests of all parties." Denver and the Front Range should "sit around a table and try to iron out some of the major issues."[22]

Since long and bitter court battles seemed to be the habitat in which Glenn Saunders thrived ("This is what I enjoyed most about the practice of law: the adversary proceeding," he acknowledged as an old man, "the vigor of a head-to-head contest."), his expression of a preference for a more congenial, less litigious way of dealing with water conflicts unavoidably evokes skepticism.[23]

And yet, whether or not those statements were heartfelt, they accurately forecast the future for Denver Water in which unilateral action would yield to negotiation, and "sitting around tables trying to iron out some of the major issues" would become, in a post-Saunders world, the major line of work for the agency's leaders. Did this mean that there had been a fundamental and institutional change? Or did it mean that a pleasant and well-tailored velvet glove had been fitted over an unyielding iron fist? Here is a riddle as challenging as the conundrum posed by the complex character of Glenn Saunders.

Was there any substance in Marc Reisner's theory that, late in life, Glenn Saunders reversed his attitude toward dams? A memorable story told by Saunders's longtime law partner, Jack Ross, calls Reisner's model into doubt. Very near the end of his life, the Water Law Section of the Colorado Bar Association notified Saunders that he had been elected "to the newly created Ancient and Honorable Order of the Water Buffalo." Though "gravely ill," Saunders "turned on his dictating machine and responded by thanking them for the honor." But he was not instantly convinced that they had made the right "choice of symbolic animal." "He said," Ross reported, "that the buffalo was merely a beast of burden, while the beaver would have been a better choice for a symbol, because it is an industrial, self-starting builder of dams." There are a number of clues that the conversion Reisner thought that he observed in Glenn G. Saunders registered just a step or two short of complete.[24]

Denver Water in the Wild Days of Yore

To ensure that your proud city grows,
You must burden its rivals with woes.
Sue till they're silly;
Attack willy-nilly;
And yield not a drop to your foes.

CHAPTER FOUR

Dealing in Diversions*

Oh, I have been often too anxious for rivers
To leave it to them to get out of their valleys.

—Robert Frost, "Too Anxious for Rivers"

To earthlings, the landscape of the moon looks even more arid than the landscape of the West looked to Zebulon Pike and Stephen Long. But as the fate of Pike and Long's prediction demonstrated, early estimates of water resources are notoriously unreliable. In the first years of the twenty-first century, "a spacecraft that NASA deliberately crashed into a permanently shadowed crater" set off "a plume containing water vapor," revealing that "the moon, once thought to be desert dry, holds a significant amount of water ice within the deep, eternally dark craters near the south pole." The implication of this discovery bore a haunting—or cheering, depending on your point of view—similarity to westward expansion: "The discoveries," as one science journalist noted, "may make it easier for humans to colonize the moon, using lunar water as a natural resource."[1] If westward expansion was to set the model for the process of drawing distant hinterlands into the control of empire, then engineers would soon be at work designing a system of diversion to bring water from the moon to the earth. But it may be that the location of lunar resources places a limit on ambition that has not, so far, characterized the development of water on earth.

In the twentieth-century West, there was little hesitation in launching plans to move water from remote locations to places of heavy settlement, even though the engineering required might have seemed nearly as improbable as transporting water from the moon. Early on, Denver Water Board members, surveyors, and staff identified the Fraser River, the Williams Fork River, and the Blue River as Denver's targeted rivers of empire. As distant as they were

* **di • ver • sion** \də-'vər-zhən, dī-, -shən\: (1) the act or an instance of diverting from a course, activity, or use; (2) something that diverts or amuses; (3) an attack or feint that draws the attention and force of an enemy from the point of the principal operation. (*Merriam-Webster's 11th Collegiate Dictionary*)

from Denver, these rivers would come to be intimately intertwined with and crucial to the well-being of the city.

Originating in the high ground of the Rockies, the three rivers began in classically beautiful mountain settings, passing through forests arrayed on scenic slopes. These areas were a perfect match to popular American definitions of landscape beauty, definitions derived from the influential writings of European romanticism. Traveling along the Blue River in 1839, Thomas J. Farnham was thrown into rapture by the mountains' "dark and stately groves of pine and balsam," "the black alder, the laurel, and honeysuckle, and a great variety of wild flowers" in "the crevices of the rocks," "the virgin snows of ages whiten[ing] the lofty summits around," and "the voice of the low murmuring rivulets." "The face of the country," Farnham went on, was "often beautiful, but oftener sublime. Vast spherical swells covered with buffalo, and wild flowering glens echoing the voices of a thousand cascades, and countless numbers of lofty peaks crowding the sky, will give perhaps a faint idea of it." On his own journey along the Blue River in 1844, explorer John C. Frémont also exclaimed over the beauty of the "country alive with buffalo," the "foaming torrents" of the river, the "high piney mountains," "the snowy ridges," and "the varied growth of flowers." With similar enthusiasm, travel writer Bayard Taylor followed the Fraser River in 1866, noting the "natural gardens" among the pines. A few days later, the area along the Williams Fork River seemed to impress Taylor even more: "The general wildness and picturesqueness of the scenery was an ample repayment for our toil."[2]

As the rivers dropped down from the higher elevations, however, their claim on beauty also took something of a fall in the minds of some visitors. Americans were slow to develop an enthusiasm for landscapes dominated by sagebrush; compared to the beauty of a soaring pine, a stubby and scrawny bush looked pretty bad, an impression that was not relieved by the sight of these far-from-noble bushes extending to the farthest horizon. Some observers, however, shifted their aesthetic standards to find value in this landscape. Reaching a zone of "sage-brush and flowers" along the Fraser River, Taylor described "a plain of silver-gray, sprinkled with a myriad minute dots of color. The odor which filled the air was so exquisite as slightly to intoxicate the senses." With sensibilities strengthened and enhanced by the exercise of making a living as a travel writer, Taylor offered a refined appreciation of this terrain: "In color, gray predominates, but a gray

Travelers who encountered the Blue River in the late nineteenth and early twentieth century often recognized the stream by its muddy tint, a telltale sign of the intense gold mining upstream near Breckenridge.

A hydraulic mine near Breckenridge around 1880. Note eroded embankments, falling water, jets of water from the monitor nozzles, sluices, fumes, a crane, and miners. This sediment would eventually wash into the Blue.

The Reliance Gold Dredging Company's Bucyrus-type hydraulic placer gold bucket-line dredge on the Swan River upstream from its junction with the Blue around 1906. Dredges like these entirely churned over the riverbed to extract any gold that may have lodged there. In this photo, Colorado School of Mines students are observing the dredge's operations.

most rare in landscape,—silvery over the sage-plains, greenish and pearly along the slopes of bunch-grass, and occasionally running into red where the soil shows through the thin vegetation."[3]

The setting of the Blue River impressed Taylor as much as the surroundings of the Fraser and Williams Fork. But this river presented a new feature: "When we caught a glimpse of the water," he wrote, "its muddy tint—the sure sign of gold-washing—showed that we had found the Blue River." Approaching the mining town of Breckenridge on the sources of the Blue, Taylor encountered "ditches, heaps of stone and gravel, and all the usual debris of gulch-mining." In a historical transformation often overlooked by those who would later oppose the disruption of streams and rivers by dams and ditches, the enterprise of mining had given the Blue River a rough time. With the detritus of hydraulic mining and the disruption of dredging, the Blue River, like the majority of Colorado's streams after the mineral rushes of the late nineteenth century, was already a river of empire, altered by human action before any water development agency could exert its power.[4]

The abundant flow of these rivers down from the high peaks, particularly in spring and early summer with the melting of the mountain snowpack, represented a consequential arrangement for the natural and human geography of the state. The melting snow predominantly flowed westward toward the Pacific Ocean, while the lion's share of the state's residents settled on the eastern side of the Continental Divide. By the middle of the twentieth century, the concentration of population and the flow of water in Colorado were not much of a match. In 1950, the US Census registered a state population of 1.3 million of which nearly 416,000 resided in the City and County of Denver and another 150,000 in the metropolitan area outside of Denver itself.*

This added up to just over 46 percent of the state population calling the Denver metropolitan area home. By contrast, the rivers of the Front Range (Rio Grande, North Platte, South Platte, and Arkansas) account for approximately 14 percent of the flows leaving the state, while the rivers west of the Continental Divide (Colorado, Dolores/San Juan/San Miguel, Gunnison, and Yampa/White/Green) account for approximately 86 percent of the flows leaving the state. Given the geographical mismatch between

*Definitions of the *Denver Metro Area* shift in different census years.

population density and the natural course of water, the pressure to transport water from the places where water was abundant to the places where people were abundant steadily increased.[5]

The resulting alignment of attitudes was, however, far from a simple opposition of a heavily populated urban Front Range pitted against an agricultural, irrigation-oriented Western Slope, rich in water. First, irrigation occurred in cities and towns as well as in rural areas, even if watering lawns and gardens seemed—and was!—very different from commercial farming. Second, commercial agriculture played a central role in the economies on both sides of the mountains; in 1939, farms on the eastern plains harvested a little more than half of the state's farm production. And third, distinctly non-agrarian enterprises like merchandising, town site speculation, and oil shale production figured in the Western Slope's hopes for water-dependent development. In other words, business leaders in the Front Range and on the Western Slope had very similar ambitions for growth and profit. And yet these *similarities* in their hopes deepened the competition and tension between the state's subregions. Meanwhile, the unsettled relations between the farmers on the eastern plains and the urbanites and suburbanites of the metropolitan area provided the Front Range with the makings of its own urban-rural divide.[6]

All of these conflicts within the state took place within a larger regional context. The Colorado River Compact of 1922 had divided the river between the Upper Basin (Wyoming, Colorado, Utah, and New Mexico) and the Lower Basin (California, Nevada, and Arizona). At the time of the compact, the states of the Lower Basin, especially California, were far ahead of the game in developing the water from the Colorado River and putting it to use. Thus, circumstances outside Colorado made disunion within the state a risky indulgence. Struggles within the state's borders made it difficult to stand in unity when Colorado's western neighbors pushed for their own interests. If Coloradans from the Front Range and Coloradans from the Western Slope fought among themselves, their common cause—developing the state's share of the Colorado River—would be weakened. Thus, division and squabbling upstream in Colorado would be a wonderful boon to the downstream states, since more water would flow to them if it were not put to use by the Upper Basin states. The more Coloradans struggled over transbasin diversions *within* the state, the more Californians would have to celebrate.[7]

From the 1930s to the 1960s, operating in this complex intrastate and interstate network of relations, Denver Water moved to a rhythm bearing a certain similarity to the archetypal western dance, the country-western two-step. Population growth provoked a nimble two-step of imposing limits on service *and* seeking a greater supply of water that would make those limits unnecessary. In the 1930s, the Denver Water Board pushed its beneficiaries to think and act on the need for conservation and reduced usage. At the same time, the board undertook an expansion of the system with the first transmountain diversion from the Fraser River through the Moffat Tunnel, as well as, in short order, a second diversion from the Williams Fork River. In a similar two-step maneuver in the 1950s, the board designated a boundary it called the Blue Line beyond which it would not provide service to suburbs *and* launched a major diversion from the Western Slope's Blue River. Asking which of these steps—limitation or expansion—met with greater favor from the local audience does not make for a very prolonged guessing game. The preference for expanding the supply clearly won the hearts and minds of the city's residents. Denverites expected the Denver Water Board to bring water from the hinterlands to the city.

When it came to worries over the adequacy and reliability of its water supply and the capacity of its infrastructure, the Denver Water Board never had a honeymoon. With the prolonged battle to merge the last two contesting companies, followed by the long and choppy campaign for the city's purchase of the private company, the water system had gone without the advantage of a steady and deliberate effort to plan for the future. Thus, from the moment of its founding, the board found itself confronting a backlog of maintenance tasks as well as a need to improve and expand its system of distribution. The first board was, as engineer-turned-historian Earl Mosley put it, instantly "confronted with demands for plant expansion," the "replacement of obsolete units," and a "backlog of deferred maintenance." As Water Board publicist and chronicler Walter Eha observed, "In 1899, the population served was 129,000; in 1919, it was 270,000, an increase of 141,000 in 20 years." Trying to provide water to their growing number of customers, the first members of the Water Board had fallen behind before they could get started.[8]

While the Denver Water Board's commissioners and engineers went after more water, they performed the requisite two-step and also sought the benefits of conservation. Very early in its institutional life, the agency was exhorting residents to avoid the overwatering of their lawns and the careless use of water in their households. In 1922, the board instituted an alternate-day water sprinkling schedule for customers. As early as 1925, Denver Water Board publications were featuring the admonition, "WATER IS PRECIOUS—SAVE IT." The board's discussion of irrigation schedules was detailed and thorough, specifying a small size for hose nozzles, determining that the water from sprinklers could not reach sidewalks, and decreeing that the penalty for noncompliance with restrictions on watering would include shutting off the water supply to the offender's property. Nonetheless, per capita rates of consumption remained high, and population growth in the 1920s continued to burden the company's water treatment capacity.[9]

By the 1930s, the idea of relying solely on water from the South Platte began to strike more and more Denverites as a precarious state of affairs. In a report in 1932, Fred Carstarphen, a consulting engineer for the Denver Water Board, declared that "it is not safe to rely upon a single stream or watershed for a metropolitan water supply." Denver would have to take up "the trans-mountain diversion of new water."[10]

Carstarphen echoed a growing chorus of city officials and water commissioners who had reached the conclusion that the future of Denver required the acquisition of water from the other side of the Continental Divide.

This line of thought gathered force from the persistent drought that burdened eastern Colorado, along with most of the Great Plains, amplifying the misery of the Great Depression. The early and mid-1930s delivered, in the summary of one group of experts, "the most widespread and longest lasting (and most famous) drought in Colorado recorded history. Severe drought started up in 1931 and peaked in 1934 and early 1935." Spells of heavy rain occurred in 1935 and 1938, but 1939 was "an extremely dry year." As the sun withered crops, hot, dry winds pulled moisture from the fields and lifted loose soil into the air. The dust surged across the land, obscuring the sun and descending upon towns as though the earth itself was swallowing them. In March 1935, a dust storm near Lamar, Colorado, "became so severe that schools were closed, automobiles traveled the highways with lights on, and nearly everyone stayed indoors." A few days later, "automobiles on highways leading into Denver were forced to pull to the side of the road, in many cases, and wait until the storm abated," while the road from Broomfield to Denver "was buried beneath inches of heavy dust." On March 26, 1934, "the worst duststorm since the US weather bureau was established in 1871 swept Denver" in midafternoon, "enveloping the entire city within 15 minutes." The storms were dreadful ordeals, stinging eyes, bringing on "dust pneumonia," silting into every crack and cranny in homes and barns, blowing away farmers' livelihoods, and robbing the soil of nutrients.[11]

In the 1930s, the threat of shortfalls and the drawn-down condition of many reservoirs put renewed energy into pleas for conservation. In 1932, state engineer M. C. Hinderlider asked "the management and owners of canals, town and city officials, to exercise every endeavor toward conserving to the greatest degree, the water which is now available and may be available before next year's demands." In the same year, after an alarming assessment of the city's water resources, Mayor George Begole and Water Board president Frederick R. Ross appealed directly to users to restrain themselves. In May 1934, the commissioners reminded consumers that "Water Is Vital!—It Is Life! Use It But Don't Ever Waste It!" By this time, however, some reservoirs were already dry.[12]

The board's conservation efforts persuaded a number of Denverites, who tried to reduce their water usage. Consumption numbers do not plot a steady downward course during the decade, but Denver did record significant reductions in water usage during December 1934, December 1935, and January 1938. And yet when spring rolled around, Denverites generally held to their pride in and dedication to the city's green expanses of lawns and gardens. Over the decades, reacting against the plains setting, middle-class residents of Denver had devoted themselves to creating an oasis in the middle of a brown expanse of native grasses. Even in the midst of the Depression and drought, one publication called Denver the City of Beautiful Lawns, and another made the doubtful claim that "Denver has more and greener lawns than any other city of its size in the world."[13]

Homeowners were so reluctant to endure brown lawns that the Water Board had to institute and strictly enforce irrigation schedules, reminding residents—even as the Dust Bowl held the nation's attention—that it was "not necessary to water lawns and shrubs every day." Even so, in 1941 the Work Projects Administration's classic

In response to the severe drought of the 1930s, the Denver Water Board urged people to conserve water even as the agency worked to develop new sources. This May 1936 advertisement on a streetcar (on Curtis Street at Fourteenth Street) reminded residents that "Water is Denver's Greatest Natural Asset—Please Don't Waste It."

guide to Colorado took notice of Denverites' unfailing commitment to green landscaping: "Novel to many is the sight of householders lovingly sprinkling their lawns and shrubbery both morning and evenings. No newspaper reports are scanned more attentively than the edicts of the Denver Water Board governing the hours of irrigation. Residents pay a flat rate based upon the number of outlets on their property; automobile owners pay an additional 50¢ quarterly." In 1959, irreverent commentator Robert L. Perkin suggested how entrenched this habit of mind had proven to be, noting that the "garden hose and the lawn sprinkler deserve a place on the city seal along with mountain sky line and soaring eagle."[14]

Tensions between the desires of residents and the constraints of the city's water supply generated discontent that found a target in the Denver Water Board. Critics looking for someone to blame for an impending water crisis knew where to point the finger. Just a little more than a decade before the onset of the Great Depression, in 1918, the board had come into existence with a chorus of affirmation from Progressive reformers. Replacing a private company, a publicly owned utility was supposed to operate with an elevated level of public accountability. To reformers, public ownership had an almost magical capacity to cleanse institutions of narrow motives and to send them out, refreshed and redeemed, to pursue the best interests of the public. And yet the fact that the commissioners on the Denver Water Board were not elected, but appointed by the mayor, left the impression that they were immune from public judgment or mayoral oversight. The drafters of the board's charter had wanted to insulate the commissioners from shifting, and sometimes even corrupting, political forces. But this effort to shield the Water Board from degrading influences created, for some, the impression that the board enjoyed a distressing excess of legal and fiscal autonomy.[15]

When drought hit the area, this apparent lack of accountability heightened some residents' anxiety over the ability of the Water Board to respond to the dilemma. The commissioners became the recipients of strident criticism, the loudest of which came from *The Denver Post.* Possessed by both an incurable distrust of the Denver Water Board and an equally incurable desire to sell papers, the *Post* seized every opportunity to accuse the agency of inefficiency and wrongdoing. The newspaper, for instance, alleged that the Water Board had not been sufficiently assertive in guarding its water rights, losing its hold on some water rights for Cherry Creek and the South Platte River. Officials, the paper's editors implied, might have purposefully surrendered "direct flow water rights so that the city would be forced into going ahead with the expensive [Moffat] tunnel project."[16]

Such allegations of mismanagement propelled an effort to override and even abolish the Water Board. The idea that Denver would be better off if an elected, not appointed, authority controlled the water supply resonated with some of Denver's residents. Proposing an amendment to the charter, a citizens' movement sought to bring the Water Board under the direct control of the mayor. *The Denver Post* supported the amendment. The *Rocky Mountain News* opposed it, claiming that the proponents simply wanted to "turn control of the water supply over to politicians and selfish private interests." Voters defeated the amendment by a margin of two to one in the spring of 1933.[17]

The majority of voters evidently felt something close to satisfaction with the first fifteen years of the Denver Water Board.

Whatever the opinions held of the water agency, there was no debate about the seriousness of the circumstances it confronted in the 1930s. "Due to deficient stream flow," the *Rocky Mountain News* reported in September 1932, "the reservoirs were unable to fill, and were completely drained in the fall of 1931." Denver's reservoirs had "great rings encircling the shores of storage reservoirs," Walter Eha wrote in 1936, that "told the story of a daily diminishing supply." As severe as the drought and the Depression might be, the Denver Water Board would not concede that nature or financial calamity might limit the supply of water available to support the city's growth. In tough times, consumers might be coaxed and exhorted to use water more judiciously, but the Water Board also took the drought as a call to increase the water supply. This effort had two primary dimensions: the maximum use of the South Platte, and then, when that river could not provide enough water for the growing city, a big transfer of water across the Continental Divide.[18]

With Cheesman Reservoir placed as the cornerstone of a comprehensive water storage system on the South Platte River, the Denver Water Board during the 1920s and 1930s expanded its capacity for storage on the river. The board looked first to acquire an existing storage facility on the river, setting its sights on Antero Reservoir. Built by a private water company in 1909 near the headwaters of the South Platte where Salt Creek comes into the South Fork of the river, Antero's storage capacity of approximately 20,000 acre-feet was just over a quarter of Cheesman Reservoir's roughly 79,000 acre-feet of capacity. While the City of Denver had been trying to buy it since 1915 (before the Water Board was even created), much of Antero's water ensured that a steady supply of water flowed through the High Line Canal, a serpentine sixty-six-mile-long irrigation ditch that had been supplying water to the agricultural fields south of Denver since 1883. The Water Board brushed aside concerns that these prior agricultural appropriations made Antero a "waterless reservoir," as an attorney for a group of farmers claimed, and finally purchased the reservoir and canal from the Lost Park Reservoir Company in 1924, adding them to the Denver water system.[19]

With the acquisition of Antero accomplished, the Denver Water Board did not rest. In 1926 they secured water rights that would allow them to build a new reservoir along the South Platte sixty miles upriver from the city. Construction on Eleven Mile Canyon Reservoir began in 1930, and when it was completed in 1932, it supplanted Cheesman as the largest reservoir in the Denver water system, capable of impounding nearly 98,000 acre-feet and sending it downstream and into the city system when needed.[20]

Even though the Denver Water Board had added the equivalent of one and a half Cheesman Reservoirs to the South Platte system in less than a decade, the commissioners could not sit complacently as the city they served continued to grow without a concession to drought or Depression. Believing that they had tapped the South Platte to full capacity, the Denver Water Board felt certain it was time to cross the Continental Divide.

The idea of bringing water from the other side of the mountains was hardly new. Other enterprises had already put the idea into practice. In 1860, miners built the Hoosier Pass Ditch, diverting Western Slope water for placer mining at the headwaters of the South Platte. By 1910, several additional, similarly small-scale ditches moved water from the Colorado Basin to the South Platte Basin. Other projects brought water from Wyoming into Colorado; by 1910, three transbasin diversions crossed the divide between the North and South Platte, moving water from the headwaters of the Laramie River into the South Platte Basin.[21]

As early as 1914, advisers to the city had promoted the acquisition of the Western Slope's water as a necessary feature of Denver's

Completed in 1932, the Eleven Mile Dam holds back nearly 98,000 acre-feet of water on the upper South Platte River as insurance against drought on the Front Range.

future prosperity. In that year, the recently created Denver Public Utilities Commission appointed a committee of engineers to assess the city's water system. "As the water for irrigation is being appropriated constantly," the engineers predicted, "the present apparent surplus on the Western Slope will soon disappear, and Denver must be early in the field as an appropriator as well as a purchaser of additional supply." One of the members of the committee referred specifically to the possibilities presented by the Fraser and the Blue Rivers on the Western Slope. The Public Utilities Commission also asked a local engineer, Ralph Meeker, "to prepare a report for the Utilities Commission on the availability of water on the Western Slope which could be used to supply Denver." Meeker's strikingly prophetic report, submitted in May 1914, explored the possibility of diversions from the Fraser, Williams Fork, and Blue Rivers on the Western Slope. "If the day is not here," Meeker wrote, "it is surely on the horizon, of the time when the City of Denver will at least have to seek a supplemental supply beyond the mountains."[22]

Very soon after the creation of the Denver Water Board, the organization began to explore possible diversion projects, focusing on the tributaries of the Colorado River, particularly the streams identified as promising by Meeker in 1914. In 1920, the board employed engineer George M. Bull to "investigate the acquisition of Western Slope water rights." Bull "began field work on Fraser River rights on July 4, 1921," Earl Mosley wrote, "while work was also being done that summer on the Williams Fork portion of the system." George Bull was an action-oriented sort of fellow. "In 1922," as James Cox wrote, "Bull made preliminary surveys for filing purposes on the Blue River and its tributaries, and filing for Blue River water was made in 1923." Right in the midst of Bull's work, a board for engineering review appointed by the Denver Water Board put forward, in August 1922, the rationale behind these actions: "It would be a very short-sighted policy to limit the growth of Denver by failing to provide a sufficient water supply for the future. The supply available from the Eastern Slope is inadequate for the future Denver. Provisions should be made at once to secure a perpetual right to all possible diversion from the Western Slope."[23]

It was crucially important that Colorado's law of prior appropriation allowed for the diversion of water from one basin to another. In Colorado water law, the key question rested on priority in timing, not geographical proximity. As James Cox put it, in declaring that water in Colorado was the property of the public and subject to appropriation, the Colorado Constitution said "nothing of eastern slope water, or western slope water." Cox extended this observation in a striking way: "An appropriator on the eastern slope has just as much right to a priority on western slope water as a western slope appropriator to a priority on eastern slope water," though this interesting reversal has yet to happen (even though proposing it would make for a fine piece of performance art). Western Slope water had traveled, over the millennia, a natural course from the Rockies to the Colorado River, just as water from the Front Range traveled to the Missouri and Mississippi Rivers, but that original route of flow carried little relevance in the legal system of allocation. The first person or organization to file on a water right, if this party put the water to "beneficial use" and satisfied other requirements, would become the senior appropriator, and the status of seniority trumped what might seem, to the unsophisticated, to be the primary rights of the residents of the claimed water's basin of origin. This recognition of the legality and legitimacy of transbasin diversion would be affirmed in the decision of the Colorado Supreme Court in 1939. In *City and County of Denver v. Sheriff et al.*, the court's decision declared that "geographical advantage does not apply to water for beneficial use."[24]

In other words, water would be allocated on the basis of the seniority of the user's right and *not* on the basis of the proximity of the user to the river of origin.

But how was the water to be moved from one basin to another? David Moffat's railroad company had already begun work on a train tunnel under the Continental Divide, which proved to offer the most viable option for transporting water. Bored under James Peak in central Colorado for a distance of 6.2 miles, the Moffat Tunnel spared trains the difficult trip over Rollins Pass and dramatically shortened the distance between Denver and Western Slope towns. Parallel to the tunnel that the trains would take, engineers had also bored a pioneer tunnel. (Pressing the telegraph key that set off the blast of dynamite that completed this tunnel was an unusually dynamic action for the legendarily lethargic President Calvin Coolidge.) Early in the tunnel's history, the Denver Water Board had recognized that this pioneer bore could be reconfigured to convey water from the Fraser River on the Western Slope into the South Platte system on the Front Range. Denver mayor

Ben Stapleton championed the idea, appointed Denver Water commissioners favorable to the project, and sought ways to secure financing for it. In 1922, the state legislature created the Moffat Tunnel Improvement District and Commission, a special governmental entity to retrofit the pioneer tunnel for water.[25]

How was a project of such a large scale to be funded and financed in times so tough that, as Stephen Leonard and Thomas Noel noted, "Denver faced bankruptcy by the end of 1933"? As with many of the public works projects that transformed the nation, the answer to that question was Franklin Roosevelt's New Deal. In October 1933, the City of Denver filed an application for Public Works Administration (PWA) money. In January 1934, Colorado voters approved a federal aid package to complete the Moffat Water Tunnel Project (an easy enough approval for Denverites to bestow since it involved no commitment to local taxes or bonds). The project met the goals of the New Deal by putting unemployed men to work while also easing anxiety about water scarcity. Under the terms of an agreement reached on November 8, 1934, with the secretary of the interior, the Denver Water Board turned over the ownership of the system to the federal government for thirty years. The government then leased the system back to the city. At the end of the thirty years, the entire project automatically became city property. The United States donated $1 million outright to the project and loaned about $2.5 million more, to be repaid from water department revenues. In all these arrangements, a "small world" phenomenon must have added an element of congeniality to federal-city negotiations: engineer George Bull, who had in the 1920s done some of the earliest work in pursuit of Denver's claims to Western Slope water, held the office of regional director of the PWA in 1933, proving to be very supportive of Denver's cause.[26]

Boosted by federal PWA money, work progressed until, on June 19, 1936, a crowd gathered to celebrate as the first water flowed through the Moffat Tunnel. While drought would continue through the decade, with the tunnel's potential to deliver nearly 100,000 acre-feet of water per year, Denverites were no longer so vulnerable to the unpredictability of the South Platte River.

In simple terms, the Moffat Tunnel Diversion System took water from the Fraser River, transported it under the Continental Divide, and sent it eastward, downhill to Denver. Simple terms, however, rarely capture major transbasin diversions like this one.

As the Moffat Tunnel neared completion in February 1936, steelworkers prepared for the installation of a concrete lining.

In its complicated geographical layout, the actual project defies any simple summary.

On the eastern side of the Rockies, the sensible strategy of using existing creek and river channels, rather than constructing a whole new network of ditches and aqueducts, was one factor inducing complexity. In the system's initial design, the water that came through Moffat Tunnel went into South Boulder Creek. Then, just above Eldorado Springs, the South Boulder diversion canal/conduit moved the water to the headwaters of Big Dry Creek. Big Dry Creek skirts around the north end of the city and drains into the South Platte near Fort Lupton, *downstream* from Denver.

Because it was downstream from Denver itself, the location of Big Dry Creek's merging with the South Platte thus called into play the far-from-obvious concept of exchange water, a way of turning to legal terms and structures to make the physical constraints of moving water around the landscape more manageable. For the first year (but only for that first year) in which the Moffat Tunnel Diversion System operated, Denver used Western Slope water entering the South Platte downstream from the city to compensate eastern plains farmers for water that the city that had withdrawn or held in reservoirs upstream.[27]

Over the next years, changes in the system added another challenge (or three or four!) to any effort at simple description. An additional portion of federal funding was secured in 1935, enabling the Water Board to reconfigure the system to take the water directly to Denver, rather than deliver it downstream from the city. Accordingly, the Moffat Tunnel Extension Unit, completed in 1937, bypassed Big Dry Creek and continued on to Ralston Creek. Ralston Reservoir then allowed the water to settle before being released into the Moffat Treatment Plant (completed in 1937), which then sent the treated water into the city's distribution system. This replaced the initial arrangements by which Western Slope water had arrived in the South Platte downstream from the city.[28]

In 1939, Denver Water Board engineers completed a second transmountain diversion, today known as Gumlick Tunnel (originally called the Jones Pass Tunnel), to bring the water of another Western Slope river to the city. This time the engineers brought water from the Williams Fork, a river flowing northwest from the Continental Divide to its confluence with the Colorado River near

On June 19, 1936, crowds celebrated the first water to flow through the Moffat Tunnel toward drought-stricken Denver.

The Denver Water Board completed Williams Fork Dam in 1938. The nearly 97,000-acre-foot reservoir stores water for Western Slope users as replacement for diversions from the Fraser and the Williams Fork Rivers.

Parshall. Built in partnership with the Denver Department of Public Works and with the financial assistance of the federal PWA, the Gumlick Tunnel moved Williams Fork water under the Continental Divide, delivering it into Clear Creek on the Front Range. Meanwhile, back on the Western Slope, the Denver Water Board built the Williams Fork Reservoir in 1938 to store water to replace flows from both the Fraser and the Williams Fork that, had they not been diverted to the Front Range, would have gone into the Colorado River.

Once the Williams Fork water was delivered into Clear Creek, the stream transported the water to its confluence with the South Platte River, north of Denver near Commerce City. This Western-Slope-originating water then supplemented the flow of the South Platte as it continued through and past Denver to fulfill obligations to downstream users, making it legitimate for the Denver Water Board to use an equivalent amount of water from upstream on the South Platte. The new infusion of water diluted Denver's sewage discharge before it reached the fields of farmers north of the city who drew irrigation water from the South Platte. This somewhat more sanitary arrangement reduced health concerns about contaminated produce from these fields, concerns that had previously led to a boycott of Colorado vegetables in neighboring states. Although the Denver Water Board did not hold jurisdiction over sewage, the benefits of diluting pollution to less-disturbing levels was a significant factor driving the diversion from the Williams Fork. As in earlier days, issues of water quality and water quantity remained closely linked.[29]

This exchange system served Denver through World War II and for more than a decade after, but by the late 1950s a new tunnel was in the works. In 1958, with new sewage-treatment arrangements on the horizon, Denver eliminated the inconvenience of exchanging Williams Fork water downstream for more storage upstream by adding the Vasquez Tunnel to the system. The new tunnel gave the diversion system an even more Rube Goldbergian quality, moving water that had recently been brought through the Gumlick Tunnel to the Front Range back under the Continental Divide to the Western Slope! Contrary to a reader's first impression, this arrangement carried considerable logic: the Vasquez Tunnel put the water (originally, remember, from the Williams Fork River) into Vasquez Creek, a tributary of the Fraser River. Thus, Williams Fork River

The Denver Water Board's Moffat Tunnel utilized the pioneer bore of the parallel Moffat Railroad Tunnel, both seen here as they appear today at their west portal near Winter Park. Today, the water tunnel is capable of delivering 100,000 acre-feet per year to Denver.

water and Fraser River water met, merged, and entered the Moffat Tunnel, traveling from there on a (comparatively!) straight shot into the Denver distribution system.

Drilling miles of tunnels through rock and laying out an extraordinary chain of canals and ditches, the designers of the Williams Fork Diversion System showed the force of engineering at its peak of ingenuity.[30]

♦♦♦

Beyond the Denver Water Board's service area, other organizations around the state matched Denver in their passion, ambition, and commitment to their locales. In the 1930s, two new organizations came into being: the Northern Colorado Water Conservancy District, along the Front Range north of Denver, and the Colorado River Water Conservation District, on the Western Slope. Led by energetic men and carrying considerable public support, these two organizations embodied the principal reason why the Denver Water Board's attempts to operate as a centralized, unilateral empire throughout the state could never be completely successful. These entities, along with other counterpart organizations in the state, pursued a range of techniques for obstructing—and occasionally joining forces with—each other's attempts to secure more water. By the mid-twentieth century, the answer to the question of who was going to acquire and hold onto the definitive power over this intricate contest was far from clear.

A reservoir on the Western Slope, little known today, would prove to be a key to this contest. Compared to the celebrity status of Lake Mead (behind Hoover Dam) and Lake Powell (behind Glen Canyon Dam), the Green Mountain Reservoir on the Blue River is far from a household name, even in households that have been directly affected by its history. In the middle of the twentieth century, however, the contest for control over this reservoir preoccupied everyone who followed struggles over water in Colorado.

How did this reservoir, which would become such a flash point for one of the biggest tussles in the state's water history, originate?

In the same era in which Denver built the Moffat Tunnel and diverted water from the Western Slope's Fraser River, the farmers and towns in the area to the north of Denver formed the Northern Colorado Water Conservancy District (Northern for short). This

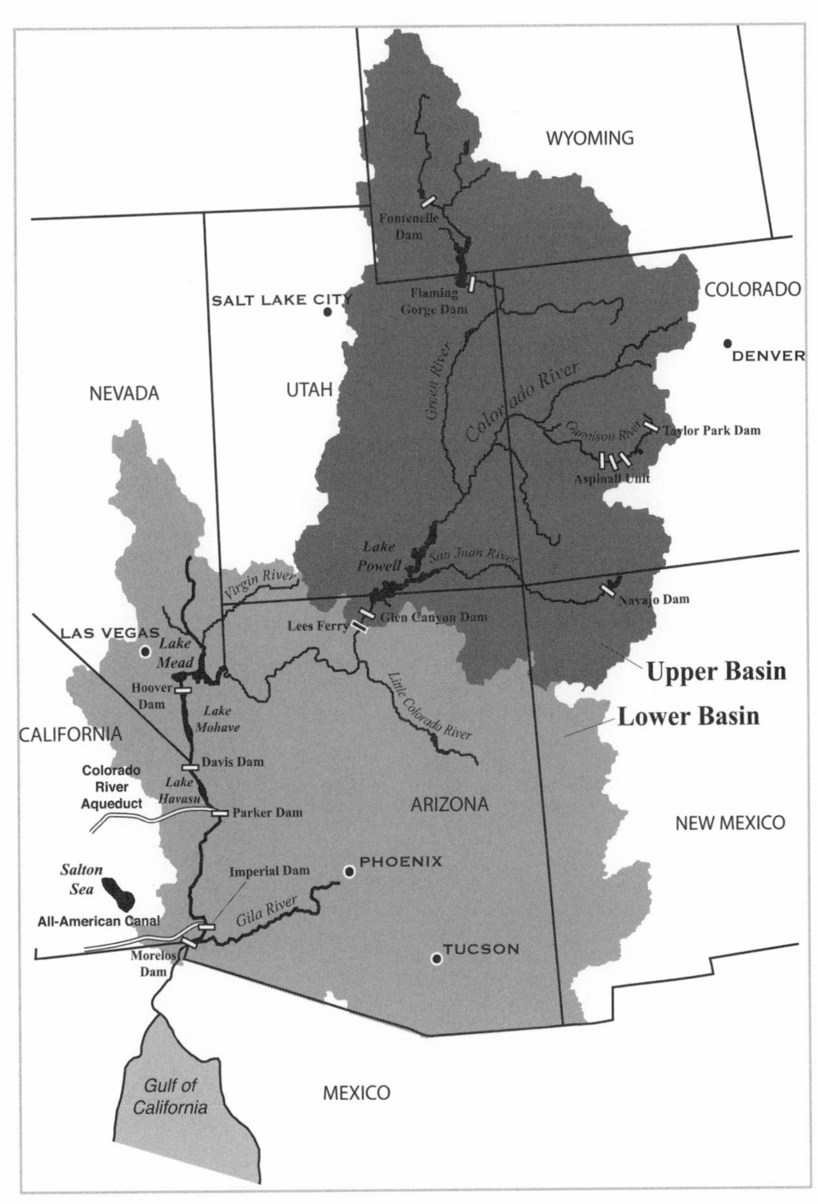

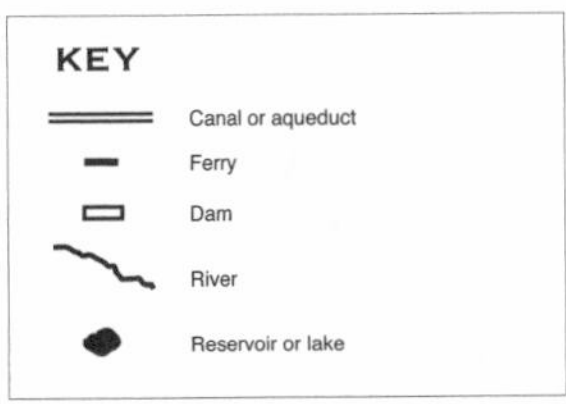

The contests for water between Colorado's Western Slope and Front Range occurred within the larger regional context of the Colorado River Compact of 1922, which divided the river's flow between the Upper Basin and Lower Basin.

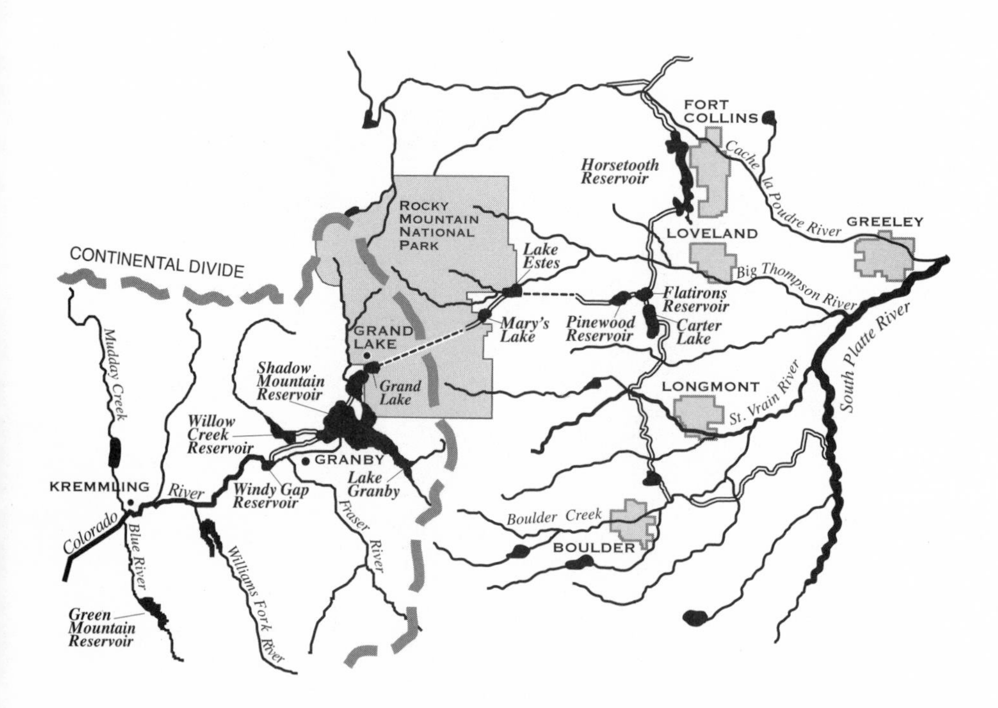
FORT COLLINS
Horsetooth Reservoir
Cache la Poudre River
GREELEY
ROCKY MOUNTAIN NATIONAL PARK
LOVELAND
CONTINENTAL DIVIDE
Lake Estes
Big Thompson River
Flatirons Reservoir
Pinewood Reservoir
Carter Lake
Mary's Lake
South Platte River
GRAND LAKE
Shadow Mountain Reservoir
Grand Lake
LONGMONT
St. Vrain River
Mudday Creek
Willow Creek Reservoir
GRANBY
Lake Granby
KREMMLING
River
Windy Gap Reservoir
Colorado
Fraser River
Boulder Creek
Blue River
Williams Fork River
BOULDER
Green Mountain Reservoir

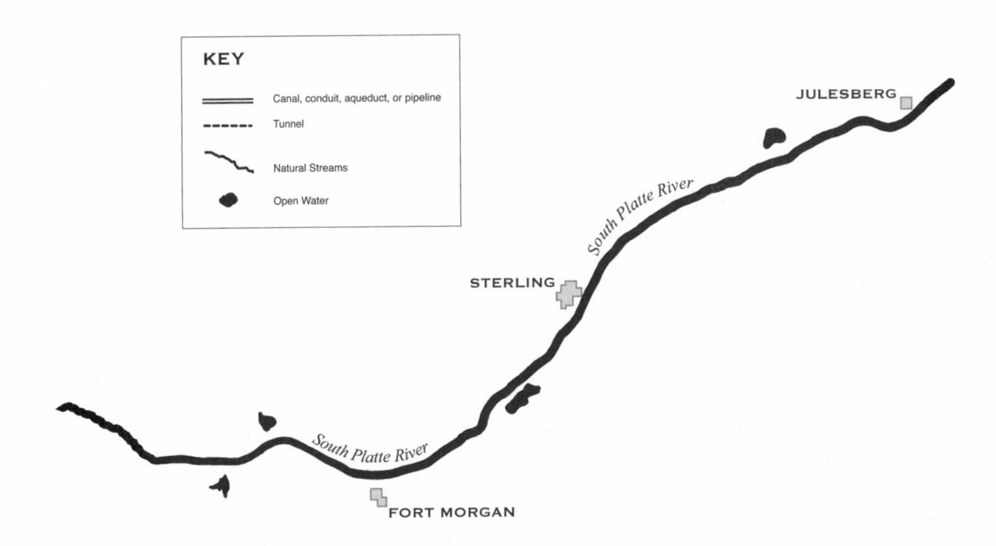

The Colorado–Big Thompson project of the Northern Colorado Water Conservancy District and the Bureau of Reclamation diverts water from the headwaters of the Colorado River to communities throughout northeastern Colorado.

new organization built a tie with the powerful federal agency the Bureau of Reclamation. The fruit of this collaboration, the Colorado–Big Thompson Project, took water from the Colorado River headwaters and diverted it through the Alva Adams Tunnel (oddly enough, passing underneath Rocky Mountain National Park) to the farms of Weld, Larimer, Morgan, and Boulder Counties, and to towns like Fort Collins, Greeley, Loveland, and Boulder.

A key part of the negotiations behind this project involved earnest efforts on the part of the leaders of Northern to conciliate key groups from the Western Slope, especially the Colorado River Water Conservation District, assuring them that there would always be an alliance between these two sections of the state and that Northern would consistently support the legitimate water claims of the Western Slope. The down-to-earth manifestation of this commitment was the construction of an underrecognized but crucially important water structure: the Green Mountain Reservoir.

Completed in 1944, the Green Mountain Reservoir stored water from the Blue River for three purposes: exchange water, compensation water, and hydropower generation. For the exchange, the reservoir stored 52,000 acre-feet of water annually to reimburse the Western Slope for water diverted out of the Colorado River Basin by Northern's Colorado–Big Thompson Project. As compensation for the loss of the diverted water, the reservoir also maintained an additional pool of 100,000 acre-feet designated for the present and future use of Western Slope entities. Finally, the release of this compensatory water was used to generate hydroelectric power, and the sale of this power helped to pay off the construction debt for the Colorado–Big Thompson Project. The federal Bureau of Reclamation, Northern's partner in this project and the owner and operator of Green Mountain Reservoir, oversaw both of these functions. And all of these arrangements—especially the purposes of Green Mountain Reservoir and the commitment of Northern to support the Western Slope's ambitions for development—were laid out in a key text known as Senate Document 80. This hard-bargained and much-cited text was an unusual treaty of alliance between two sections of a state whose interests and ambitions could just as easily have made them enemies. Representative Edward T. Taylor of Glenwood Springs, a longtime advocate for the Western Slope and chairman of the powerful House Appropriations Committee, threatened to "kick the slats out of the whole business" by withholding funding for the Colorado–Big Thompson Project unless the Western Slope got its fair share in the agreement. Taylor insisted that for every acre-foot diverted under the Continental Divide, an equal amount of compensatory storage must be provided for water to be used on the Western Slope. He held fast to his position through years of political horse-trading, and Senate Document 80 enshrined the important concept of compensatory storage in Colorado's water law.[31]

In the 1930s, the Denver Water Board's transmountain diversions through Moffat Tunnel and the Gumlick Tunnel and Northern's operation of the Colorado–Big Thompson system proceeded with surprisingly little friction between the two projects and their sponsoring agencies. While the water resources of Colorado could not have struck anyone as unlimited, the parallel development projects of the 1930s did not pit the people of the Denver area against the people of northern Colorado. And yet, with the creation of these two separate projects, there were hints of a future in which no one agency could impose its will without the consent—or the defeat—of another.

These clues to the future were, however, easy to miss and far from definitive. By the end of the 1930s, only people afflicted with madly overactive imaginations could have glimpsed the world ahead, in which unending tugs-of-war and collisions between local, place-based groups and organizations would come to be the new normal. On the contrary, lessons drawn from the successful incorporation of water from the Fraser River and the Williams Fork River into the ambitions of Denver ratified one of humanity's most lasting platitudes: *Where there's a will, there's a way.*

Perhaps the most significant lesson, as convincing at the time as it would prove to be misleading in the future, concerned the federal government. By 1940, it seemed self-evident that the primary federal role in these matters was to provide funding for the enormous water projects that would benefit particular locales. While it took enterprise and determination to secure this funding, several widely shared premises predisposed both the legislative and the executive branches of the federal government to respond favorably to local initiatives: (1) the prosperity of the nation could not be separated from the prosperity of the country's regions and sections; (2) big projects thus contributed both to local *and* national well-being; and (3) public works provided the federal government with an appropriate and necessary mechanism for creating jobs in an economic depression. Beyond the provision of funding, a federal agency might on occasion grant a right-of-way across public land or cede a few acres for a reservoir. But the idea that laws, regulations, and administrative actions of the federal government might, over the course of a couple of decades, turn out to be principal obstacles to big water projects would have struck any sensible person in the 1930s as a fevered and implausible fantasy. The story of the Denver Water Board in the 1930s immerses us in an era that, for twenty-first-century readers, will seem exotic in its chipper and enthusiastic (though still not unanimous) attitude toward spending on behalf of what was understood to be the public good. In the 1930s, where there was a will, there not only was a way—there was also federal funding.

The first round of diversion of water from the Western Slope to the Front Range built confidence, not humility. The leaders of Denver Water had every reason to take on the future with an attitude that had been thoroughly validated by achievement. The agency's engineers had triumphed over probability and over the Rockies themselves, moving water from one side of the mountains to the other. Its lawyers had acted as an advance guard for the engineers, clearing the way to federal support and to vindication of prior appropriation and transbasin diversion in the Colorado Supreme Court. Encountering little opposition, Denver Water had conquered geography and extended its power far into the hinterland. Reaching for the moon, the leaders of the agency had reason to conclude, was a strategy that worked. The water pouring into Denver from the Fraser and Williams Fork Rivers offered inarguable testimony that confident and odds-defying ambition delivered results. No longer "leaving it to [rivers] to get out of their valleys,"* the Denver Water Board had redesigned and reconfigured the way water traveled over the surface—and under the surface—of the planet Earth.

* When quoting Robert Frost on a topic involving the American West, it is always useful to remind readers that the iconic New England poet was born in California. It is also important to note that there is no evidence that Frost intended his poem "Too Anxious for Rivers" to address transbasin diversions!

The Labyrinths of Water Law

If you study Blue River Decrees,
You will soon grow weak in the knees.
Say you're fit as a fiddle—
This stipulated riddle
Will soon make you totter and wheeze.

Chapter Five

A Horrifying Jigsaw Puzzle*

The Uncertain Course of the Rivers of Empire

In 2010, Jim Lochhead, manager of the Denver Water Board, chose colorful, and not exactly neutral, language to characterize his agency's history. "Denver Water was certainly at one point the evil empire," he said. "We live in a much different time today."[1]

The image of a city and its agencies operating as an empire—evil or otherwise—is both powerful and lasting. In 1933, in an influential book called *The Rise of the City, 1878–1898,* Arthur Meier Schlesinger either coined the phrase *urban imperialism,* or gave it new visibility. Writing of the dominance of Chicago over its hinterland, Schlesinger declared, "No better example could be found of what one contemporary called 'urban imperialism,'" though he did not give a name to that contemporary. Schlesinger was more precise in pinning down the term itself: "As the world's greatest corn, cattle and timber market, Chicago completely dominated the Mississippi Valley and, to some degree, the farther West as well." Chicago's urban empire centralized the flow of the resources of a vast hinterland, assigning the ambitions of the residents of that hinterland to a place in the queue well behind the ambitions of Chicago.[2]

Environmental historian Donald Worster contributed another memorable term to the lexicon of western water history with his influential book *Rivers of Empire.* In the management of western water and particularly in the operations of the federal Bureau of Reclamation, Worster found a pattern in which "power becomes faceless and impersonal, so much so in fact that many are unaware that it exists." In a hydraulic society, a centralized elite ruled over both human beings and nature, Worster believed: "Accepting the authority of engineers, scientists, economists, and bureaucrats along with the power of capital, the common people became a

---

* Reporter Bert Hanna, in *The Denver Post* in 1963, describing the issues, actions, court cases, and negotiations that formed the context for the Blue River Decree of 1964. (Bert Hanna, "Trial Delay Raises Hopes for Out of Court Dillon Settlement," *The Denver Post,* December 17, 1963.)

herd." This book, first published in 1985, glued the concept of empire to the history of western water development.[3]

In 1992, historian William Cronon provided fresh meaning to Schlesinger's idea of urban empire by connecting the history of the city to the history of the environment, tracing Chicago's acquisition and distribution of natural resources from rural hinterlands. "From the heart of the city," Cronon wrote, "the frontier history of the Great West looks to be a story of metropolitan expansion, of the growing incursions of a market economy into ever more distant landscapes and communities."[4]

For those on the receiving end of urban exercises of power, the accuracy of characterizing cities as empires has seemed indisputable. From the perspectives of residents of the hinterlands, the arrogant cities decreed, dominated, and exercised their unilateral will, indifferent to the resentment of their subjects. Episodes of resistance and rebellion only provided these empires with the opportunity to remind the rebels how weak they were and how unyielding were the agents of empire. Subdued, mastered, and unilaterally ruled, the rivers of empire and the residents of the colonized hinterland surrendered and submitted to urban imperialism.

The concept of urban imperialism might seem suited to the City of Denver and to the Denver Water Board. Consider the call to arms delivered in 1903 by the president of the Denver Chamber of Commerce, Meyer Friedman: "In the relentless war for commercial supremacy fought by cities, states, and nations, only that community can be victorious which brings to bear upon its rivals, a well-equipped and persistent army of citizens." Friedman's remark conveyed the spirit of many of his fellow business leaders; Denver had taken for itself the regal title of Queen City of the Plains and, in the words of historians Stephen Leonard and Thomas Noel, "strove to conquer the Rockies." It had been the resources of "its big economic backyard," Leonard and Noel wrote, that "assured Denver's success." Long after its founding, "the city's role as a processing, transportation, trade, and service center remained constant." Historian Kathleen Brosnan's appraisal coincides with Leonard and Noel's. The city of Denver, she explains, "dominated regional urban growth and environmental change. In what might be called a system of urban primogeniture, Denverites emerged early and never let go. The city became the center for directing people, resources, information, and capital into and out of the region." Similarly, after its creation in 1918, and especially after its Western Slope diversions of the 1930s, Denver Water Board's story also seemed to be a tale of expanding power over the city's Front Range neighbors and over the hinterland of the Western Slope. By the truism "If the shoe fits, wear it," Denver Water seemed fully outfitted for a major excursion into urban imperialism.[5]

And yet, at second glance, Denver Water's history comes well-supplied with stories of the agency having to negotiate with competitors, to adapt to the growing authority of the federal government, to learn to work within constraints of nature, money, and political accident, and to recognize that many of its actions would trigger resistance and rebellion from opponents who did not defer to the idea of an empire that gave no room to opposition or resistance. As Leonard and Noel repeatedly observe, Denverites themselves also knew that "theirs was a colonial economy," often dancing "on strings pulled in New York boardrooms." However it might try to dominate its hinterland, the city itself "suffered from its lack of industry, distance from large markets, and the power exercised over its destiny by eastern financiers." Brosnan ratifies this judgment: "As the Denver region moved closer to the center of the nation's economy, both physically through improved transportation and communication systems and more abstractly through financial institutions and capital investments, its entrepreneurs discovered that the control they initially exercised had dissipated." In the same way, if the Denver Water Board achieved any episodes of uncontested exercise of power, these episodes were not of long duration.[6]

Rather than steering an imperial course to the future, Denver Water often seemed to be pulled and pushed into action by an increase in population driven by distant changes. In many areas of the American West, World War II brought a boom in industry and population. The summation of Denver's changes, offered by Leonard and Noel, deserves full quotation: "World War II triggered a tremendous transformation in Denver. Massive federal spending, an influx of newcomers, and a pent-up demand for new cars and new housing unavailable during the war led to a boom that changed a drowsy provincial city into a sprawling metropolis." The sprawl moved laterally: "In the 1950s and 1960s Denver was not growing up anywhere nearly as fast as it was growing out," Leonard and Noel note, leaving the relationship between new suburbs and the city chronically unsettled.[7]

By the 1950s, the Denver Water Board looked out at this sprawling and thirsty metropolitan area—now home to 566,000 people and many trees—and felt that they needed to limit their service area until they could provide more water.

By the early 1950s, an expanding population caused Denver's water managers to return to another performance of the two-step maneuver of the 1930s: try to limit consumption while also seeking new sources of water. This time, the board went beyond trying to persuade consumers to avoid waste, taking the big step of limiting the population the department was obligated to serve. In August 1951, as growth throughout the metropolitan area edged the water supply toward its projected limits, board commissioners drew boundaries around the area in which the Denver Water Board would be able, with its already operating projects, to supply water. The Blue Line enclosed an area 114 square miles, of which 56 square miles were the City and County of Denver.[8]

Areas outside the Blue Line would not receive service from Denver Water. Individuals and companies would have to think hard before locating homes or businesses outside the Blue Line, since water service would be uncertain. Developers, officials at lending institutions, merchants, and aspiring homeowners would invest in the outside districts only at their peril. For those outside the city limits but within the Blue Line, Denver Water would offer contracts for service that charged much higher rates than those inside the city. These contracts, moreover, operated only on a year-to-year basis, allowing Denver Water to adjust the amount of water it would deliver according to fluctuations in the unpredictable annual snowpack.[9]

It was not that the previous arrangements had been marked by either great generosity or flexibility. Under a policy established in 1948, the board asserted the right to a "reasonable return" on the investment needed to provide water to out-of-city customers. The most striking consequence was this: all out-of-city customers had to be metered and charged according to the scale of their use, while in-city customers enjoyed flat-rate billing, so that higher consumption did not result in a higher bill. Finally, and most troubling to suburban communities, the Denver Water Board reserved the right to stop outside service when it was in the "best interest" of the City of Denver.[10]

The Denver City Charter's mandate to provide water, first and foremost, to residents of Denver explained each of these conditions. There was, finally, no avoiding the charter's insistence that the agency's first obligation was "supplying the City and County of Denver and its inhabitants with water for all uses and purposes." Any leases

The Denver City Charter is clear about the Water Board's top priority: "Supplying the City and County of Denver and its inhabitants with water for all uses and purposes." And by the 1950s, Denverites of all ages had come to enjoy numerous uses and purposes for water, such as these children playing at a pond in City Park.

to supply water outside those limits must be "subject to the future needs and requirements of the City and County and its inhabitants."[11]

Denver citizens had good reason to appreciate the vigilance with which the Water Board watched over their interests. But officials of cities like Aurora, on Denver's eastern flank, could not see much compatibility between these conditions and their own well-being. In the decades after World War II, Aurora's leaders wanted to embrace their own opportunities for growth, and they did not want to leave their city's destiny dependent on the plans and preferences of the Denver Water Board. In 1949, Aurora secured its own Western Slope water rights on Homestake Creek and five of its tributaries. Joining forces with the City of Colorado Springs, south of Denver, Aurora completed the Homestake Water Collection and Storage System in the late 1960s. The Denver Water Board hoped to participate in this project, but Aurora rejected any such overture. Aurora had become a competitor with the Water Board, setting a trend in the metropolitan area away from centralized regional management and toward a fragmentation of the water supply among many small and competitive organizations. From a point of view situated on the Western Slope, it might have been possible, in the last half of the twentieth century, to perceive a united, imperial opponent called the Front Range. But the conflicting interests of Denver and its suburbs created fault lines and canyons that fractured that supposed monolith, and the Water Board found itself on increasingly tenuous footing in its efforts to act as the coordinating and orchestrating force on the Front Range.[12]

Unwilling to comply with the Denver Water Board's requirements or to accept the Blue Line's arrangements of insiders and outsiders, other suburban governments, such as Englewood, to the south of Denver, followed Aurora's inclination for independence. One approach led to the establishment of special government districts. By 1962, at least twenty-six of those special water districts were scattered around the periphery of Denver, as well as another thirty-three water and sanitation districts. Many of these special districts, particularly the smaller ones that did not have the resources to build their own systems, depended wholly or partially on the Denver Water Board. Others aspired, on a smaller scale, to the success achieved by Denver. But their late start put them at a competitive disadvantage, with Denver far ahead in infrastructure and its portfolio of senior water rights.[13]

The growth of suburbs was raising enormous questions for the future of the Front Range, while also calling into question the ostensible imperial power of Denver to direct or limit the processes of development and land use. In sharp-edged commentary, Leonard and Noel pointed out characteristics in the mind-set of suburban residents and their officials: "On this suburban frontier . . . people continue to pursue the American dream—capitalizing on nature's bounty, demanding individual freedoms and opportunities, and resisting government regulations and taxes." Denver's neighboring counties steered by the values of privatism, trying "to perpetuate the mobility, individual freedoms, property rights, and rural ideas of the Old West in new subdivisions." Even if the Western Slope seemed to be the Denver Water Board's obvious rival and opponent, the individualistic suburbs in the metropolitan area represented as substantial a challenge to the agency's efforts to centralize its power.[14]

In 1386, Geoffrey Chaucer used the word *suburb* in *The Canterbury Tales.* Scholars have identified this as the second time the word appeared in English. In *The Canterbury Tales,* Chaucer did not, however, set any precedents in suburban boosting. "In the suburbs of a toun," Chaucer wrote, lurked "robbours" and "theves" in "fereful residence." By the mid-twentieth century, city leaders all over the nation were struggling with the ungainly challenge of suburbs that demanded services for which their residents were not always willing to pay. Coordinating the governance of a city and its suburbs could have tempted knowledgeable city officials to consider adopting the Chaucerian tradition in characterizing suburban qualities. As urban historian Jon C. Teaford summarized this challenge, "Bands of suburban municipalities ringed many central cities, and in a single metropolitan area scores of municipalities shared governing power." In the Denver area, as in many metropolitan areas, the result "was overlapping authority, conflicting policies, and yet another barrier to the unity of the metropolitan population." This national trend was the result of purposeful choice: as Martin Melosi put it, "Suburbs themselves added to social, economic, and political fragmentation by sometimes resisting consolidation with the central cities." This situation presented many domains of vexation, with water supply proving to be particularly fruitful of friction.[15]

Patterns of precipitation were, in any case, producing conditions that discouraged generosity among Front Range communities. The years between 1950 and 1956 were dry (with the brief exception of a snowy winter in 1951–52), dropping water supplies to uncomfortably low levels. Even when reservoirs were refilled in 1957 by one of the wettest years in recorded history, the Denver Water Board had found in the drought an unnerving reminder that foresighted action could never give way to complacency.[16]

Since the drought hit the whole state, every Colorado locale received a similar message of scarcity, making it seem urgent to secure more water for the future. Plans to increase Denver's water supply led to a seemingly paradoxical shift in the Water Board's stance toward out-of-city customers. Anticipating expensive improvements to its system, the agency looked for opportunity to increase its revenues. The Blue Line had turned into an obstacle to Denver's best interests, constraining its ability to bring in those revenues. In a 1959 letter to the board, Denver Water manager Robert Millar wrote frankly of the effects of the Blue Line on metropolitan development and on the relationship between Denver Water and the suburbs:

> The Blue Line policy of the Board is so well known that many who expect to apply for the water have refrained from making those applications as long as a fixed geographic boundary exists. . . . [T]here is strong emotional resentment against Denver's attempts to plan and regulate the lives of people outside city limits.[17]

As Millar observed, the relationship between Denver Water and some neighboring communities had curdled, even if the Blue Line policy and the terms imposed on out-of-city customers had made sense in the context of the Water Board's charter responsibilities and the uncertainty of water supplies.

The chances for the coordinated development of a metropolitan water supply, bringing city and suburbs into a sense of their shared interests, had taken a plunge. Years later, assessing this era in Denver Water's history, historian J. Gordon Milliken reflected a common judgment when he remarked that the Blue Line policy had "led to fragmentation of the metropolitan water supply system, competition for water supply sources, competitive pressures to serve new customers, unwise suburban growth dependent on water supplies of dubious reliability, and destructive political feuds between Denver and its suburbs."[18] Milliken's judgment carried important freight, but it stopped short of recognizing and respecting the important experiment that ran from 1951 to 1960 (the year that the Blue Line was lifted). The Blue Line had raised two very important and persistent questions: To what degree did Denver Water's advantages in rights and infrastructure give it a corresponding power (whether or not the commissioners chose to exercise it) to shape and direct settlement and growth in its vicinity? And even if the Denver Water Board presented its neighbors with an opportunity to reappraise their expectations of an ever-expanding water supply, sufficient to support every local dream of growth and expansion, would the neighbors reduce their demands and accept any limits? Without answering those big questions about the future of Front Range growth, the Blue Line simply made the Denver Water Department seem even more imperial to its suburban neighbors.

In earlier times, when officials had presumed that Denver would annex satellite towns and cities as they emerged and entered the Water Board's territory of service, the charter's mandate to serve residents of Denver created comparatively little friction. In the first half of the century, Denver had annexed many contiguous communities and taken on their public properties, legal obligations, and debts. Metropolitan officials and planners had every reason to believe that this would continue. The Home Rule Amendment of 1904 (also called the Rush Amendment, and properly Article XX) provided not only that school districts of annexed areas would merge with Denver's school district and that Denver would assume all school bonds, indebtedness, and obligations; it also made it possible to extend to those communities the security of Denver's dependable water supply.[19]

In the context of drought and tight supplies in the 1950s, the city council reinforced and seconded the board's Blue Line policy by adopting a more conservative annexation policy. To discourage annexation, Denver passed a city ordinance in 1956 that required annexed areas to pay a fee of $2,000 per acre of residential land for capital improvements to the water system. This fee effectively halted new annexations and also the extension of water supplies to new communities.[20] When communities pursued self-governance separate from the City of Denver, the mandate of the charter restrained and even prohibited the Denver Water Board from acting as a centralized authority providing water for the whole metropolitan region.

By 1960, having lifted the Blue Line restrictions, the board had resumed the old policy of extending water service to communities that met its requirements. They did so under the auspices of a new charter, adopted in 1959, that cast off the old requirement that water-service contracts outside of the city were limited to one-year terms and instead allowed the board to negotiate permanent contracts with users beyond the city limits. The new charter specified that the board's priorities had not changed—its first responsibility was still "to provide an adequate supply of water to the people of Denver" at rates "as low as good service will permit"—but the creation of perpetual lease arrangements shifted the dynamics of the Water Board's relationships with outside communities.

With the Blue Line erased and the new charter in place, the City of Denver again looked to absorb contiguous territory in the neighboring Adams, Arapahoe, and Jefferson Counties. However, some suburbs at this time began to conclude that annexation to Denver was not necessarily in their interests, suspecting that Denver sometimes used its advantage in water to force annexations. Suburban officials feared that the City of Denver would raid their tax bases for its own needs, especially for public education. Some retailers resisted annexation because of Denver's 3 percent sales tax. Business owners also wanted to avoid paying Denver's relatively high property tax, and county governments did not want to lose property tax revenues by having their most vital areas reassigned to Denver. In a classic denunciation of Denver as empire, written to state legislators in 1961, George Creamer, an attorney who fought on behalf of property owners in the community of Glendale against Denver annexations, declared that "we know of no basis for annexation by Denver save its own insatiable desire for territorial expansion, tax and revenue increase, and economic dominance."[21]

⸫

Following its familiar two-step course of action, the Denver Water Board had actively sought new sources of water even as it was limiting its service by observing the Blue Line. With remarkable accuracy, Denver Water planners in 1953 predicted that Denver Water would have to supply 725,000 people by 1963, and "in excess of one million" by the early twenty-first century. "With no new water," Denver Water official Earl Mosley told a reporter, "we'll have to halt the metropolitan area's growth and expansion by 1963." Given the time needed to build the projects to bring new water into the system, the grace period seemed alarmingly short. With no direct authority to restrict the movement of people into its service area nor to demand that they change their expectations for a properly green landscape, Denver Water's leaders felt compelled to extend the reach of the system of water collection further into the Western Slope. As a 1953 Water Board report explained, "Since the South Platte River supply has already been developed to capacity, the needs of the next ten years must be met by prompt expansion of available facilities located on the headwaters of the Colorado River."[22]

Declarations of the need to expand the Denver Water Board's domain stirred up Colorado's internal sectionalism. To residents of the Western Slope, Denver's further ambitions for acquiring water from the Colorado River Basin were disquieting. But on the Front Range, already facing drought and a growing population, a variety of voters, candidates, agencies, utilities, and nonprofit organizations found the Water Board's call for more development to be urgent and convincing. In 1955, Denver voters overwhelmingly approved a $75 million bond measure for capital improvements to tap Western Slope water sources. In 1959, another successful bond issue added $40 million to the package.[23] Equipped with these bonds, the Denver Water Board moved into its next phase of transmountain diversions.

Why was the Denver Water Board asking for more Western Slope water even with the certainty of riling up their fellow Coloradans on the other side of the Divide? First, the Denver City Charter unambiguously identified the agency's obligation to provide the City and County of Denver with adequate water. Second, Denver Water Board planners rigorously calculated how much water their collection systems could produce and projected how much water their customers would demand; those calculations and projections identified a narrow margin that left little room for miscalculation and no room for lethargy. Third, the agency carried an institutional memory of the trying times of the 1930s, when drought fueled a fear of a looming water famine. The drought of the 1950s reconfirmed and reactivated the concerns of the 1930s. And the fourth factor was the combination of energy, ingenuity, persistence, and determination that went under the name Glenn G. Saunders, lead counsel for the Denver Water Board (readers should review chapter 3 if, inexplicably, they skipped it). With these four consequential factors in play, Denver Water was going to choose action over passivity.

Anticipating the future meant focusing on the Blue River, a stream flowing down from the Tenmile Range through valleys of conifer forests and merging with the Colorado River in arid sagebrush steppe near Kremmling. In the waters of the Blue, Denver Water planners saw an opportunity to divert to the Front Range a flow of water that would nearly double the system's holdings.[24]

Denver's claim on the Blue River originated long before the Water Board tried to put it to use, hearkening back to the surveys commissioned by the Denver Public Utilities Commission in 1914, before the Water Board had even been created. The possibilities identified in these earliest reports were elaborated in the 1920s, when the board sent consulting engineer George Bull to survey Summit County's watersheds for likely opportunities for diversion to Denver. Bull identified possible locations for dams and diversion structures and even filed claims on behalf of the city to water rights on several streams, including the Blue. In the next decades, the Great Depression, World War II, and the shortage of men and materials forced the Water Board to scale back its ambitions for development. The agency conducted no field surveys to follow up on Bull's claims until the 1940s. That delay in establishing the board's definite interest in the Blue had big consequences, leaving the date that would be assigned to Denver's rights open to dispute. Thus, the board's effort to assert and to act on its claim for Blue River water rights launched a protracted, tangled struggle with Western Slope interests.[25]

In 1942, Denver Water filed a request in the state District Court for Summit County for adjudication (that is, for a definitive allocation and affirmation of a claimed water right) of its Blue River rights, to be heard in the courtroom of Judge William Luby. Opposition came quickly into play as a consortium of Western Slope water districts and providers disputed Denver's claim. They were joined and strengthened in their opposition by the federal government's Bureau of Reclamation.[26]

What brought Reclamation into this fray? Remember the Northern Colorado Water Conservancy District's big diversion, the Colorado–Big Thompson Project. A key feature of that project—the first component of the project to be built, in fact—was the

The timing and intentions of the Denver Water Board's interest in the Blue River would become central questions in the legal wrangling over Dillon Reservoir's water rights. This photograph shows a group of men identified as a triangulation party working on the Blue River Tunnel Survey in September 1921.

152,000 acre-foot Green Mountain Reservoir, completed in 1943 to store water from the Blue River. Under the agreement negotiated between the water users of northeastern Colorado and the Western Slope, the purposes of the Green Mountain Reservoir were (and are), first, to provide replacement water in exchange for the water redirected from the Western Slope under the mountains to northeastern Colorado and, second, to compensate the Western Slope for the water taken out of the basin with a 100,000 acre-foot supply newly earmarked for present and future uses. This compensatory water within the reservoir was often referred to as the "power pool" because its release was used to generate electricity, the reservoir's third purpose, supplying power as well as revenue that was used to pay off the construction debt for the whole project.

The Green Mountain Reservoir was downriver from the new site of diversion proposed by the Denver Water Board. Thus, because Denver claimed 1914 as its priority date for the Blue, Denver's suit threatened the 1935 priority date of Green Mountain Reservoir. A reliable flow of water from the Blue River was essential to Reclamation's plans for the reservoir, to fill it and thus to supply the downstream users and also to drive the hydroelectric power plant. Reclamation argued that the Denver Water Board's dam would prevent the annual filling of Green Mountain Reservoir and would thereby interfere with the bureau's ability to deliver on its obligations.[27]

By Colorado's doctrine of prior appropriation, the plan to acquire and move water from the Western Slope would be justified and viable if Denver Water could prove the validity of its rights in court *and* establish an early priority date for them. Rights with earlier dates were senior rights, and senior rights prevailed over later, junior rights, regardless of any claims to agrarian virtue, pioneer values, urban sophistication, or cosmopolitan civilization that any claimant might make. Prior appropriation law also made it clear that a mountain range poses no legal impediment to the appropriation, transport, and use of water.[28]

While Denver's request for adjudication lingered in Summit County district court, a very complicated tale of litigation over rights to the Blue River and its tributaries began its twists and turns. The Bureau of Reclamation, which had first joined Western Slope water districts in their suit in state court, withdrew from the suit brought by the water districts and on June 10, 1949, filed a separate claim in federal district court to affirm its right to Colorado River waters over the claims of Denver and other Front Range communities (Colorado Springs and Englewood, which also claimed Western Slope waters for themselves; similarly, the South Platte Water Users Association claimed standing). As readers are surely realizing, the parties converging in various courtrooms for litigation over the Blue River made for a complicated cast of characters.[29*]

In 1952, ten years after the filing of the Denver Water Board's initial request for adjudication in state court, Judge William Luby ruled that Denver's claim to Blue River water rights was legitimate. To that degree, Denver was the winner. But Judge Luby's next finding reduced the scale of that victory considerably. The Denver Water Board had asked for a priority date of 1914 for those rights. The judge rejected that request and assigned the rights a priority date of 1946. The agency had failed to demonstrate a substantial and sustained intention to develop those waters, he said, until the summer of 1946, when they had begun work on the portal of what would become the Roberts Tunnel. The designation of the later priority date meant that the Denver Water Board had secured only about half of the water from the Blue River that it wanted.[30]

Denver Water Board lead counsel Glenn Saunders was not in the habit of accepting defeat, whether partial or complete. Two months after Judge Luby's decision, Saunders, along with attorneys for Colorado Springs, appealed the decision to the Colorado Supreme Court. As he waited for the Court's decision, Saunders also argued his case before the public. "Denver's Water Crisis Not Over, Expert Warns," the headline in *The Denver Post* read on October 1, 1954. With a drought under way, Saunders cautioned that it "might take several years to get the city's water supply up to normal." But if the city's plans for the Blue River succeeded, "its water problem will be solved for all time." Dismissing the opposition from the Western Slope, Saunders declared that there was "enough water in Colorado for both East and West Slopes." Moreover, he asked all Coloradans to recognize that "we all prosper if Denver prospers."[31]

On October 18, 1954, the Colorado Supreme Court upheld

---

* These parties included the US government (in particular, the Bureau of Reclamation in the Department of the Interior), the Northern Colorado Water Conservancy District, the Colorado River Water Conservation District, the Palisade Irrigation District, the City and County of Denver, the City of Englewood, the City of Colorado Springs, and the South Platte Water Users Association.

the lower court's decision with a 4–3 decision *against* Denver. In an alignment full of meaning for the state's history, the judges split along geographic lines: the three from Denver sided with the Denver Water Board, and the four from outside the city proved to be the majority. In the next weeks, the Colorado Supreme Court denied several motions from Denver requesting a rehearing.[32]

"Court Kills Hope for Water," the headline on October 19 shouted. The *Rocky Mountain News* characterized the decision as "probably the most devastating legal blow ever delivered against this city and its possible development." If it stood, "it would mean an absolute end to the growth of Denver within boundaries already clearly drawn," and it would "make impossible the provision of a single additional drop of water outside the already defined limits."[33] The city's residents were unsettled and alarmed. Real estate developers suspended plans for large residential projects. The famed Del Webb abandoned a plan to build a six-thousand-home community. Newspaper columnists predicted that the city, if unable to expand its population and economy, would stagnate and decline. The statement that the Colorado Supreme Court's "decision would clamp a lid on Denver's growth by 1963" became a common feature of public commentary, accompanied by the short-term threat that "Denverites may be told to quit watering entirely for the rest of the year."[34]

"It is hard to say what the total effect of the ruling is or what our next step will be," Glenn Saunders said in mid-December 1954, concluding with an entirely characteristic confidence: "One thing we know—Denver must have that water from the Blue." Appearing undaunted by this setback in the state courts, "Saunders gave the impression of a man with an ace in the hole," one reporter noted, quoting the attorney's promise that "We'll come up with something" to get Denver the water it needed.[35]

At this juncture, events took an improbable turn away from litigation and toward negotiation. Despite the Colorado Supreme Court's ruling, the US District Court for the District of Colorado had agreed to hear the suits over the Blue River in federal court.*

---

* The two cases from the state District Court for Summit County (Civil Actions No. 1805 and No. 1806) had been moved to the US District Court for the District of Colorado in May of 1955, where they were consolidated with Civil Action No. 2782, the original action launched by the US government. The date for the trial of these consolidated cases was then set for October 5, 1955, laying out an unmistakable deadline for compromise.

Dismay over the prospect of another prolonged battle in the courts played a part in the multiple decisions that added up to a shift. Just as important, the larger context of regional water issues placed a premium on unity within the state of Colorado. When the Colorado River Compact divided the river's flow between the Upper Basin and Lower Basin in 1922, the Lower Basin had been much more active than the Upper Basin in developing water; unless it got to work fast, the Upper Basin risked being left behind and ending up with a less than equitable share. The bill creating the Colorado River Storage Project, with a package of projects for the Upper Basin, was stalled in Congress. If the state of Colorado remained fractured and divided, it would be nearly impossible to get this bill passed. The necessary demonstration of unity was, unmistakably, a compromise on the Blue River. Another push toward negotiation came from a very high level indeed, as President Dwight Eisenhower made it known that he thought compromise was in order.[36]

A pending court date of October 5, 1955, offered an incentive to negotiation and set a clear framework for action. In September 1955, lawyers from the Denver Water Board, the Bureau of Reclamation, the Northern Colorado Water Conservancy District, and the Colorado River Water Conservation District met to try to reach an out-of-court settlement. Each organization could send an attorney and an adviser. One change in the usual cast of characters may have carried particular consequence: Glenn Saunders was ill, and Harold Roberts, a man with a more conciliatory temperament, took his place. Just weeks before, as historian Dan Tyler put it, "conciliation [had] appeared as unattainable as ever." But, as the trial date bore down upon them, the participants forged a compromise on October 4, presented this "stipulation" to the presiding federal judge on the morning of October 5 in place of making opening statements, and on October 12, Judge William Knous signed the agreement and gave it official standing as the Blue River consent decree of 1955.[37]

The settlement recognized Senate Document 80—the agreement of collaboration between the Northern Colorado Water Conservancy District and the Western Slope that made possible the Colorado–Big Thompson Project—as a governing document that set the parameters of the Blue River agreement. This meant that when it came to the 52,000 acre-feet held in Green Mountain Reservoir as exchange water to replace the Colorado–Big Thompson's diversions, Front Range claims were junior to the claims of Western Slope users. The decree reaffirmed the priority dates handed down by Judge Luby, giving the Reclamation Bureau's Colorado–Big Thompson Project (and thus the Green Mountain Reservoir as well) a priority date of August 1, 1935, senior to the Denver Water Board's date of 1946. Denver and Colorado Springs agreed to release diverted water back to Green Mountain Reservoir if it failed to fill in a given year. The two cities pledged to recompense the Bureau of Reclamation for any loss of revenues from electrical power it could not generate because of their diversions from the Blue River. Despite these conditions—the key for the Denver Water Board—Denver and Colorado Springs were given the right to divert Blue River water, if they stayed within their priority, to the Front Range for municipal uses. The Western Slope litigants did not prevail with their demand for the metering of all Denver customers or for the limitation that Blue River water be used only by customers within the city limits, but they did win an agreement that the Denver Water Board would exercise due diligence in reusing return flows of water imported from western Colorado.[38]

The Blue River Decree seemed to settle the major arguments among Denver, the Western Slope, and the Bureau of Reclamation. One newspaper did make the important observation that, while leaders from both the Eastern and Western Slopes saw the accord as "a stepping stone to fuller use of water," the "Western Slope leaders were somewhat more reserved in their acclaim." In its most immediate result, the decree revived the feasibility of Denver Water's Blue River Diversion Project and paved the way for Congress's passage of the Colorado River Storage Project on March 28, 1956 (the act included, by reference, the terms of the Blue River Decree). The peace achieved by the decree would not prove to have much in the way of durability, but the process that produced the decree still signaled the arrival of a new era. Rather than an omnipotent empire, the Denver Water Board was now only one participant in a big group of well-represented contenders competing for the same supply of water.

Figuring out how to operate in this new era had tested—and would continue to test—the willingness to negotiate and the flexibility of an organization that had grown accustomed, in its early years, to achieving its goals by strength and force of will. Even if representatives of the Denver Water Board sometimes played the part of urban imperialists, their institution operated in a tangle of contesting interests and external authorities. Adopting the posture and speaking the scripted lines of empire was proving to be a very different matter from actually exercising imperial power.[39]

♦♦♦

By the time the federal court approved the Blue River Decree, the Denver Water Board was well launched on building the first major element of the Blue River water collection system: the tunnel that would convey water under the mountains from the base of the planned reservoir that would store Blue River water, to near Grant, Colorado, on the Eastern Slope. The name Harold D. Roberts Tunnel would honor the attorney who stepped in for the ailing Glenn Saunders to negotiate the 1955 Blue River Decree. At 23.3 miles long, the Roberts Tunnel marked another engineering triumph as one of the world's longest water tunnels. The system's second component would be a storage dam located at the confluence of the Blue River, the Snake River, and Tenmile Creek, at the town site of Dillon. Water from Dillon Reservoir would then travel to the Front Range through the Roberts Tunnel.[40]

Even as construction got under way, the Denver Water Board continued its campaign, conducted around the negotiating table and in the courtroom, to secure as much water as possible from the Blue River. The 1955 decree seemed to have settled disputes over the river, but the agreement contained ambiguities that were open to reinterpretation, and the restless mind of Glenn Saunders was made for this mental sport. He made one more try to secure more of the water stored in Green Mountain Reservoir. He found his inspiration in 1959 when the paper company Crown Zellerbach asked the Bureau of Reclamation for Green Mountain Reservoir water for use in a plant to be located near Kremmling. If a paper-producing company could request water from the Bureau of Reclamation, Saunders thought, why couldn't a city? Green Mountain water, Saunders suggested, "may find its highest and best use for the people of Colorado in Eastern Colorado."[41]

Federal officials and Western Slope advocates felt differently. The Department of the Interior, the home of the Bureau of Reclamation, gave "a flat denial to Denver's claim." Saunders's response was predictable: the official from Interior who issued this

Even as litigation over diverting the Blue River dragged on, the Denver Water Board pressed ahead with the construction of the diversion tunnel, which would come to be called Roberts Tunnel. The work spanned more than two decades, a strategic pace that helped secure Denver's water rights claim during the long legal dispute. The photo shows a prominent inspection party, including Denver mayor Quigg Newton, at the tunnel's east portal in August 1949.

response, Saunders said, "leaves us no alternative but to ask the court to correct him." At a meeting in Washington, DC, convened by officials of the Department of the Interior in November 1959, Saunders was unyielding. Prominent Western Slope advocate John Barnard, attorney for the Colorado River Water Conservation District, voiced the suspicions of many when he argued that Denver did not actually need Blue River water for its own use but wanted to sell it to downstream farmers in the South Platte Basin, in order to generate revenue to pay off the expensive expansion of its system. The city had, Barnard claimed, "secretly planned an ambitious, comprehensive, and entirely ruthless scheme."[42]

In a manner that did not seem to qualify for Barnard's adverb *secretly*, Saunders and the city leadership pressed ahead on every front, continuing to assert an interest in the water stored at Reclamation's Green Mountain Reservoir and moving forward with their own dam at Dillon. By the fall of 1960, the peace achieved with the Blue River Decree of 1955 was thoroughly frayed. Headlines declaring "City's New Water Plan Draws West Slope Ire" and "Inter-Slope Water Warfare Erupts Again" proclaimed the reopening of the battle.[43]

One dimension of the new round of controversy arose from the Denver Water Board's hope to insert a role for the Williams Fork Reservoir into this already very complicated system. This reservoir did *not* store water for diversion to the Front Range but, instead, provided the Western Slope with water to replace what had been sent out of the Colorado River Basin, from the Fraser and Williams Fork Rivers to the Front Range. Originally built in 1938 to replace water diverted through the Gumlick Tunnel, the Williams Fork Reservoir had been expanded in 1959 to account for the increased diversion due to the Vasquez Tunnel's completion, and now Denver Water wanted to expand it once again so that the reservoir would also provide replacement water for water diverted from the Blue River.[44]

While the use of the Williams Fork Reservoir was one point of contention, the core of the conflict hinged on the Denver Water Board's position that "its proposed domestic use of Blue River water should take precedence over the use of Blue River water by the United States for power purposes." The Bureau of Reclamation insisted that it needed water to maintain a 100,000-acre-foot power pool to produce electricity in Green Mountain Reservoir's turbines.

Saunders and the Water Board had anticipated this objection with the enlargement of the Williams Fork Reservoir in the late 1950s; at that time, they had installed a generating plant over the objections of the local water districts. This generating power plant made it possible for Saunders to pledge that Denver would make up for the loss of power generated at Green Mountain either with Williams Fork–produced kilowatts or with monetary compensation. But Reclamation sided with the Western Slope, where its facilities and water holdings were intended to spur development; indeed, the proposed Crown Zellerbach plant was one example of the kind of development that Reclamation wanted for the Colorado River Basin. Thus, the Department of the Interior refused the Denver Water Board's initiative to secure water stored in Green Mountain Reservoir for Front Range purposes.[45]

After the failed effort at negotiation in Washington, DC, a coalition of western Colorado water districts, including the Colorado River Water Conservation District, responded to Denver Water's claims. On October 4, 1960, the coalition filed a petition in federal court to prevent Denver and Colorado Springs from claiming any share of the water in Green Mountain Reservoir. Moreover, Denver's intention to use Williams Fork Reservoir to provide replacement water would, the petitioners contended, pose a serious obstacle to another possible Bureau of Reclamation enterprise (the Parshall Project), a prospective dam on the Williams Fork that would provide water to Western Slope ranchers and farmers.[46]

Pending court actions did not slow down construction of the Roberts Tunnel, completed in 1962, nor did it delay the building of Dillon Dam. Despite the lawsuits, the Denver Water Board's leaders assumed that eventual decisions and rulings would allow for storage and diversion of Blue River water, and workers began constructing the massive earthen berm that would hold back the river. Perhaps surprisingly, the fact that Dillon Reservoir would flood the site of the existing town of Dillon did not arouse much public concern. The small town was nearly nomadic, having already gone through three relocations—the first for a better situation along the Snake River, and the next two to accommodate railroad lines—and most Coloradans (those who did not call Dillon home) were not particularly upset about one more move. There was some irony in the coinciding of Denver Water's two parallel challenges: arriving at a plan to move the town *and* figuring out how to house

The old town of Dillon stood at the three-way confluence of the Blue River, the Snake River, and Tenmile Creek. This view, taken before 1960, looks south from near the site that would become the west abutment of Dillon Dam.

The Denver Water Board relocated the town of Dillon in order to make way for Dillon Reservoir. This photo shows the Dillon church being moved to its new site in June 1962.

350 construction workers near their workplace. Along with nearly two hundred living residents, the preparation of the site required the relocation of the town's cemetery. In 1961, after condemnation proceedings completed the acquisition of houses and properties, Dillon residents moved to their current town site directly south of the area that would be submerged by the reservoir.[47]

In July 1963, Western Slope water districts tried one more legal action (a petition to amend their 1960 petition!) to prohibit the filling of Dillon Reservoir. Denver's plans to begin filling the reservoir, the city's opponents contended, would undercut Green Mountain Reservoir's senior rights to the Blue River and thus take the water needed for Colorado–Big Thompson replacement and for power generation. Neither the Bureau of Reclamation nor the Western Slope water districts found anything to like in Denver's plan to replace Blue River water with Williams Fork water; the result would be an equal amount of water downstream in the Colorado River, but the water would bypass Green Mountain Reservoir and its power turbines. The Western Slope's legal efforts evoked the usual rhetorical pep from Glenn Saunders. As *The Denver Post* paraphrased, "in the opinion of Glenn Saunders," this litigation added another layer to "harassments of long standing to impede and obstruct 'for selfish motives' Denver's development of water resource for growth and increasing needs of the metropolitan area."[48]

The pace and tenor of events began to bear some resemblance to a classic Western showdown. Secretary of the interior Stewart Udall, on behalf of Interior's Bureau of Reclamation, ordered Denver not to finish the dam until the federal litigation was resolved. The Water Board did not give an inch. In "the face of an order issued by Udall not to use Dillon Dam," Denver Water manager William Miller and attorney Glenn Saunders pushed ahead toward the completion of the dam as scheduled, ever certain that the city's future—not to mention millions of dollars in bonds—hung in the balance.[49]

By the beginning of September 1963, the Denver Water Board was facing down a coalition of opposing Western Slope interests and the federal government. With the critical moment approaching, Saunders's theatricality, along with the joy he took in a no-holds-barred public feud, was on conspicuous display. "No federal court or department can keep Denver from acquiring water," he declaimed. "That dam is being paid for 100 percent by Denver," and, in Saunders's interpretation, Denver was in full compliance with the 1955 Blue River Decree. Federal officials had no standing to intervene in the operation of Dillon Dam; "it's none of their damn business, and I'd like to see them try it," Saunders said. "We're not violating any decree of court." While Denver's leaders earnestly prepared for the future, Saunders said, "These people in Washington sit on their fat fannies" (this phrasing surely missed its target with Stewart Udall, one of the most physically fit public officials in American history).[50]

On September 3, 1963, the Denver Water Board stood its ground and defied the Department of the Interior as crews began to close Dillon Dam and impound the waters of the Blue River. Contrary to Interior's orders, Dillon Reservoir began to fill.[51]

Saunders remained defiant, claiming full legitimacy for Denver's actions within the scope of the Blue River Decree of 1955: "We think we have complied with the decree with the greatest care." Federal attorneys expressed their disagreement in a motion filed in district court on September 6, charging that the Denver Water Board's plan to fill Dillon Reservoir in fact violated the decree by interfering with Reclamation's obligation to provide replacement water out of Green Mountain Reservoir for its Colorado–Big Thompson diversion. The government's attorneys argued that storage of Blue River water in Dillon Reservoir constituted "direct and major invasion of the prior rights of the United States." A number of Western Slope organizations soon asked to be included in the federal government's motion.[52]

On the Western Slope, observers looked at Denver Water's defiance of Interior and saw just one more demonstration of the city's arrogance. Articles in the Western Slope's *Grand Junction Sentinel* in this era sizzled with anger, with Glenn Saunders as the principal target. Denver, the editors wrote in September, was "financially and morally bankrupt," guided by leaders who "don't care about right and wrong. They care nothing about the rights of others just so Denver's needs are taken care of. Glenn Saunders, in riding roughshod over everyone, is a fair example of all Denver's official family." They returned to the complaint the following month: "As long as Saunders runs the Denver Water show on the theory that might makes right and to the devil with anyone who disagrees there will be no chance of unity" within the state of Colorado. The negative reviews of Saunders's performance were not entirely limited to the Western Slope, either: an editorial in the *Denver Democrat* urged

Denver's mayor to fire Saunders, who had put in place "the jagged rocks of hatred, envy, and jealousy that characterize Denver's relationships with the rest of the state." Saunders, his opponents said, shifted tactics and arguments in an expedient and unpredictable way; "trying to pin him down," wrote William H. Nelson in the *Sentinel*, "is like trying to step on a glob of mercury."[53]

And yet, shifting to the point of view of the Denver Water Board, the agency had spent more than $70 million to bore a twenty-three-mile tunnel, build a reservoir with a capacity of 250,000 acre-feet, and relocate an entire town and its surrounding highways, thus rendering the Interior order not only too late, but beside the point.[54] From this angle, high-handed Washington, DC, authority should not stand in the way of the foresight of the city. While the lively rhetoric condemning Denver's (and particularly Saunders's) imperial stubbornness catches our attention, it is important to note a provision of the 1922 Colorado River Compact that seemed to validate Saunders's argument. "Water of the Colorado River System," the compact declares, "may be impounded and used for the generation of electrical power, but such impounding and use shall be subservient to the use and consumption of such water for agricultural and domestic purposes and shall not interfere with or prevent use for such dominant purposes."[55] In arguing that the 100,000 acre-feet held in Green Mountain Reservoir for the generation of electricity should be made available for the domestic purposes of Denver, Saunders's thinking was aligned with the hierarchy of uses put forward by the compact. But he was also calling into question the widely accepted assumption (one that he usually shared) that prior appropriation, as the law of the state, prevailed over an interstate compact.

Judge Alfred A. Arraj of the US District Court of the District of Colorado set April 13, 1964 (initially December 9, 1963, then given a ninety-day postponement) as the date to begin hearing arguments from the federal government and the Western Slope against the Denver Water Board. With the court date on the horizon, the litigants began to find reasons to consider an out-of-court settlement, realizing that court cases were likely to go on for an inconveniently long time.[56]

Even with episodes of grandstanding and bluster, the Denver Water Board's leaders kept a door open to compromise. In a diplomatic overture, Saunders asked assistant attorney general Ramsey

Clark for a parley on the issue, and the two met in Washington in September 1963. Back in Colorado, Saunders experimented with a different tone and style of public expression that may have surprised his critics: "Both Denver and the Western Slope would benefit materially in improved relations, if we could all sit around a table and try to iron out some of the major issues," he said. "It would mean greater harmony and both could work together to develop Colorado's remaining water resources for the benefit of all Colorado." In an effort to court such harmony, the Water Board offered an olive branch by discontinuing the storage of water in Dillon Reservoir until the litigation was resolved. In early 1964, newspapers began cautiously reporting that such a resolution might be in the works, noting "the possibility of an out-of-court settlement of a controversy that resembles a horrifying jigsaw puzzle."[57]

The approach of the scheduled trial date of April 13, 1964, galvanized the negotiators. Two days before, on April 11, headlines announced "Denver Water Board OK's Blue River Pact," though with a telling additional subheadline: "Western Slope Approval Awaited." Denver had made a major concession in giving up any claim on Green Mountain Reservoir, but the Western Slope interests still withheld their approval. The negotiators were now engaged in serious edgework. Without an agreement in place, the long-anticipated trial began on Monday, April 13. On that first day, Judge Arraj "left no doubt that he was dissatisfied" with the Western Slope's case, "and felt attorneys should try to agree on certain key issues." Speaking with a bluntness that could make an attorney's hair stand on end, Judge Arraj told Frank Delaney, the attorney for the Colorado River Water Conservation District, that he should cut back on overheated rhetoric and "come into this court and tell me precisely what it is you want. Until you're ready to do so, I think we have heard enough for today." And, showing equity in his forthrightness, the judge spoke with equal harshness to Glenn Saunders. "Mr. Saunders . . . let me make a suggestion. There's a conference room upstairs. All of you participants go up there and sit down at the table. When you have come to some agreement, you come back and tell me. I'll be glad to listen." He then continued the case until Wednesday morning, April 15.[58]

On Monday afternoon and during the day Tuesday, an intense discussion occurred among the attorneys for the various parties. Delaney, representing the Colorado River Water Conservation

Completed in 1963, the expansive 254,000-acre-foot Dillon Reservoir is the largest storage facility in the Denver Water supply system. It was also, at the time of its construction, the most intensely litigated.

District, assumed his place as the one dissenter. But by Thursday, April 16, Delaney had conceded, and Judge Arraj filed a consent decree, "agreed to in a closed-door meeting of attorneys representing the Federal Government and Eastern and Western Slope water interests." Glenn Saunders cheerfully observed that the "most important thing to come out of this entire proceeding" would be "harmonious relations" between water interests in eastern and western Colorado.[59]

The Second Blue River Decree "provided that neither Denver nor Colorado Springs had any right, title or interest in Green Mountain Reservoir," a significant victory for the Bureau of Reclamation, the Northern Colorado Water Conservancy District, and, to a degree, Western Slope interests. In return for this concession, Denver secured the right to substitute water from the Williams Fork System for the water it diverted to the Front Range from Dillon Reservoir. As long as Green Mountain Reservoir was able to fill, senior rights downstream were satisfied, the Denver Water Board relinquished all claims on Green Mountain Reservoir's water, and the board either replaced or paid for any power that Green Mountain could not produce because of these arrangements, Denver could store water in Dillon Reservoir. The implementation of the agreement was subject to the approval and supervision of the secretary of the interior, installing a significant federal presence in these intrastate relations. The decree did not take a position on the anticipated conflict between Denver's Williams Fork Reservoir and the prospective Bureau of Reclamation Parshall Project.[60]

On July 17, 1964, what *The Denver Post* reporter had christened "a horrifying jigsaw puzzle," the long and tangled story of maneuverings, negotiations, initiatives, and counterinitiatives turned into the material reality of water traveling from Dillon Reservoir through Roberts Tunnel and into the city of Denver. "The new supply," declared one newspaper reporter, "has rescued the city from what could have been a critical water shortage." The Denver Water Department had released some water "to fulfill obligations to Green Mountain Reservoir," satisfying the Bureau of Reclamation and the Department of the Interior. Glenn Saunders drew the improbable conclusion "that the arrangement marks a new era of improved relationships between Denver and the Western Slope."[61]

♦♦♦

And what, in heaven's name, is the moral to this maddeningly complicated story?

Did Glenn Saunders, representing the Denver Water Board in every sense of the word, actually "ride roughshod over everyone," as the editors of the *Grand Junction Sentinel* had charged? Did he get his way, albeit by a route that came closer to a zigzagging broken-field run than a line drive? His attempt to secure water from the Green Mountain Reservoir had been thoroughly stymied, although securing an abundant flow of water from the Blue River surely softened the pain of defeat. Did it still make sense to place Denver Water in the category of urban imperialism, if empire periodically had to surrender unilateral decision making, submit to the determinations of umpires and referees, and make the best of a negotiated settlement?

One thing was certain: the playing field for control over Colorado's water had gotten crowded and congested. Rivals and competitors—a variety of Western Slope interests ranging from ranchers and farmers to oil shale developers, the Northern Colorado Water Conservancy District, and the farmers downstream from Denver on the eastern plains—were impossible to silence, control, or avoid. The suburbs of the metropolitan area were, on their own, proving to be a sufficiently rowdy group of subalterns. And, one level above this whole boisterous and contentious cluster, the federal government was emerging more and more as an external force—a fountain of funds and, in a trend that was just getting under way, also a fountain of regulations and restrictions.

The Denver Water Board, in the judgment of one sharp-eyed policy scholar, was confronting an increasing incompatibility between the persistence of its "core tasks" and the arrival of a new set of "unwanted functions," many of them involving the management of increasing public involvement and the whole process of strategic negotiation. These functions required a capacity to navigate through a social and political labyrinth of growing complexity. For all its strengths, the Denver Water Department entered this world with considerable vulnerability and even weakness. "The water commissioners," Brian Ellison wrote, "viewed their core tasks in the starkest engineering and business terms." They saw "storing, treating, and delivering water to the citizens of Denver and leasing unneeded water to the suburbs" as "a technical problem that was dependent on the services of engineers and lawyers." A form of urban imperialism steered by engineers and lawyers was certain to mean that this particular ship of state was going to have a tough time charting its course, avoiding rocks and reefs, and reaching its intended destination.[62]

Extremely off course in his own voyaging, Lemuel Gulliver awoke to find himself in the company of human creatures who were less than six inches tall. It might have seemed that this difference in scale meant that the power was all on his side. Gulliver, however, soon discovered that even miniature opponents could seriously constrain his actions. "I attempted to rise, but was not able to stir," he said. "I found my arms and legs were strongly fastened on each side to the ground; and my hair, which was long and thick, tied down in the same manner. I likewise felt several slender ligatures across my body, from my armpits to my thighs." When he pulled his left arm loose, his situation did not improve. "In an instant I felt above a hundred arrows discharged on my left hand, which pricked me like so many needles," causing Gulliver to groan "with grief and pain," and to conclude that "it was the most prudent method to lie still." The Lilliputians may have been tiny, but like the first one Gulliver saw—"with a bow and arrow in his hands, and a quiver at his back"—the Lilliputians were not exactly pacifists.[63]

By 1964, the Denver Water Department's leadership was tied up in its own version of Gulliver's troubles. It was big, strong, and yet remarkably constrained in its ability to move vigorously and purposefully. Urban imperialism was proving to be as difficult for the city to practice as it was easy for the hinterland to condemn. But Gulliver had an option that Denver Water did not have: he could decide that "it was the most prudent method to lie still" and thus avoid further friction with the Lilliputians and their arrows. Denver Water had no such option. In 1963, just as Dillon Reservoir was close to completion, and just before the Roberts Tunnel would begin delivering Blue River water to Denver, the engineering firm Black & Veatch submitted a comprehensive review of the water system. James Lee Cox summed up its findings for the future: the firm declared that "present facilities will be inadequate by 1979."[64]

With sixteen years to go before the next projected crisis, the Denver Water Department did not have the option to choose prudent repose. There would be no rest for this weary if not wicked empire. According to the well-established customs of the organization, anticipating the threat of scarcity and securing the increased supply that would neutralize the threat, it was time for Gulliver to get up and get braced for the next barrage of arrows.

The Post–World War II Metropolitan Landscape

Throughout the American West,
The suburbs have made us all stressed.
They have eaten up farms,
Set off fiscal alarms,
And given the cities no rest.

Chapter Six

No Country for Old Habits

Foothills, Two Forks, and the Revision of the Future

A person that started in to carry a cat home by the tail was gitting knowledge that was always going to be useful to him, and warn't ever going to grow dim or doubtful.

—Mark Twain, *Tom Sawyer Abroad*[1]

Apollo was a high-performance god. One of his most impressive accomplishments was his design of error-free prophecy in ancient Greece. Responding to a question from a suppliant who wanted to know the future, Apollo would communicate through a set of priestesses who would shout and shriek in an incomprehensible manner. A priest would translate these outbursts into cryptic pronouncements arranged in hexameters. The suppliants would then go off to interpret these predictions. If their interpretation turned out to match the future, then the oracle's success at prophecy was immediately evident. If the future, however, took a different turn, then the fault lay not in the oracle, but in the inadequate interpretation of the prediction. When, for instance, the oracle seemed to tell the ruler Croesus that he should invade Persia because this would lead to the downfall of a great empire, Croesus made a bad assumption about *which* empire would take this fall. The messages of the oracle, according to one description, "were always open to interpretation, and often signaled dual and opposing meanings." The temple at Delphi was destroyed in 373 BCE, not by a grudge-holding recipient of a misleading prediction, but by a poorly foretold earthquake. Rebuilt, it was definitively destroyed several centuries later by the decree of a Roman emperor who seems to have been a risk-taking sort of ruler.[2]

A millennium or two later, most human beings are reduced to finding their clues to the future in the events unfolding around them. It is a little unnerving to see that events communicate in a manner rather like Apollo's priestesses: they carry multiple meanings, they require strenuous interpretation, and a good share of the time they are insolubly bewildering. Sometimes they baffle humans by proceeding sneakily through time, unnoticed by those whose destinies they will affect; other times, they confuse people with the equivalent of unintelligible shouts. Events are particularly treacherous when it comes to their unwillingness to be forthright in announcing major change. To take the example closest at hand, when an organization that has consistently gotten its way encounters an obstacle, it is as *reasonable* to take that obstacle to be a temporary setback as it is to recognize it as a bellwether of enormous and irreversible change.

In hindsight, it is entirely clear that by the 1980s Denver Water had received a generous supply of hints and warnings that its old habits were not going to continue to work. Return to the early 1980s, however, and a different interpretation—that the agency could master new challenges with strategy and persistence, and thus had no need to change its custom—looked like an equally reasonable conclusion. It is not hard to imagine why Denver Water found the second approach considerably more appealing.

💧💧💧

In the years 1955 and 1956, two big events with seemingly opposite meanings occurred in western water history. In the much-less-publicized event, the 1955 completion of Denver Water's Gross Dam, the expectations of business as usual were solidly validated, and the patterns and trends of the recent past continued to flow in familiar and comfortable channels. In the far more famous event, the 1956 decision not to build a dam at Echo Park within the boundaries of Dinosaur National Monument, the well-established expectation that large dams would continue to be built went sailing off a cliff and landed in a heap. For alert observers in the mid-1950s, depending on which event was at the center of their attention, it would have been reasonable to conclude that everything was changing in western water management and just as reasonable to conclude that nothing much was changing. In mid-twentieth-century Denver, the epidemic of uncertain vision that afflicted many who claimed to see the future would have caused even the best ophthalmologist to throw in the towel.

Let us begin with the event that seemed to confirm that the patterns of the past would move smoothly into the future. In 1955, the same year as the first Blue River Decree, the Denver Water Department completed the construction of the most expensive reservoir in its system to that date. Without anything like the controversy that dogged its counterpart on the Blue River, Gross Reservoir—designed to provide storage on the Front Range for water diverted from the Western Slope—was christened with "dignity and beauty" (as the Water Board happily reported about its own party) by Miss Sharon Kay Ritchie, Miss Colorado and future Miss America, who ceremoniously smashed a bottle of Western Slope water against the concrete dam.[3] People who attended the Gross Dam ceremony with Miss Colorado would have been perfectly justified in thinking that the building of dams on the Front Range would occur in an entirely congenial and ruckus-free manner, even if diversions from the Western Slope had become the occasion for legal tussles.

Nonetheless, those present at Gross Dam's christening could not miss the fact that their cheerful ceremony rested on a foundation of human tragedy. A survivor, Harry Artemis, described the calamity that took place four years earlier. On August 24, 1951, Artemis was "working 280 feet above the floor of the canyon, drilling blasting holes in the face of one wall" and "suspended by 60 feet of strong rope tied above me to the rim of the canyon." A storm raced in that afternoon, and before he and other workers could take refuge under a rock ledge, a lightning strike ignited "32 tons of dynamite on both walls" in "a series of ear-splitting explosions that bore through the canyon's belly like a bomb." Artemis and others landed on the canyon floor. "It looked like Korea, only worse," he remembered, drawing a comparison to the gruesome battlefields that appeared in the news as the dam was under construction. "Blood streaked the rocks. Dazed men were yelling, probing—looking for the dead and injured."[4] Of the 250 men who had been at work on the canyon walls when the lightning struck, 9 were dead among the rubble. Stories of this intensity are not uncommon in the origins of the undertakings that we have veiled and cloaked with the dull term *infrastructure.* For most Americans who benefit from those systems, the sedative of amnesia has been used to relieve the discomfort that might be delivered by the stories of infrastructure's martyrs.

Exempted from any comparable amnesia, paradoxically enough, was the Echo Park Dam, made famous by the fact that it was never

Denver Water dedicated Gross Dam on August 2, 1955, without any of the controversy that dogged the dams at Dillon or Echo Park. The highlight of the ceremony, according to a report from the Water Board, was Sharon Kay Ritchie, Miss Colorado and future Miss America, who smashed a bottle of Western Slope water against the dam "with dignity and beauty."

built and thus never brought injury or death to workers. Formally proposed by the Bureau of Reclamation in 1946, the Colorado River Storage Project included a dam and reservoir in a remote area near the border of Colorado and Utah. The principal goal of the Colorado River Storage Project was to hold water in the Upper Basin of the Colorado—to guarantee that the Upper Basin could honor its Colorado River Compact obligations to deliver 7.5 million acre-feet to the Lower Basin every year and to hold water for use in the Upper Basin itself. This particular dam would flood an area within a designated national monument under the management of the National Park Service. The dinosaur tracks and relics at Dinosaur National Monument would not be affected by the dam, but a beautiful area within the monument—Echo Park—would be flooded. For the many organizations that united to oppose the dam, the principle at stake was the recognition of the priority carried by federally designated land preservation. A national park or a national monument should be exempt, by this principle, from serving as a dam site.

The fight over Echo Park is a widely recognized watershed event in American environmental history. As Mark Harvey, the leading historian of this episode, observes, before this battle the Bureau of Reclamation (and, one might add, other agencies developing water on a large scale) "could not easily be challenged by those who lacked technical knowledge of irrigation, hydropower, and cost-benefit ratios." The dam's opponents, led by the Sierra Club's David Brower, knew they could not win their campaign by praising the beauty of the site they wanted to protect. Those who wanted to preserve Echo Park had to develop "an understanding of evaporation rates, the legal complexities of the Colorado River compact, and the economics of large water and power projects." Since Congress would make the final decision, the best route to success was to give senators and representatives from the North, South, and Midwest reasons to question the economic and engineering rationales for the dam.[5]

In 1956, after an extraordinary lobbying effort by preservationists, the law creating the Colorado River Storage Project did not authorize the Echo Park Dam. In exchange for stopping the project at Echo Park, environmentalists agreed not to oppose a large dam at Glen Canyon on the Colorado River. This trade-off complicates the calculus of victory in a final assessment of the Echo Park controversy. Still, Harvey sees the defeat of the dam as "a great symbolic victory for the sanctity of the national park system." It was also a victory that rippled with real consequences for water agencies, as environmental historian William Cronon has explained: "The struggle to stop the dam at Echo Park would become a defining moment in the emergence of a new post-war environmental politics . . . From this moment on, dam-builders would gradually find themselves in ever more defensive positions, so much so that many of their most hoped-for projects eventually became untenable."[6]

If the defeat of the Echo Park Dam represented the defining moment, after which dam builders would be in ever more defensive positions, that moment did not shake the complacency of the leaders of Denver Water. As is often the way with large-scale changes, the rising power of what would be called environmentalism was easy to see in hindsight but hard to detect in 1956. Moreover, the central issue of the Echo Park controversy was, in a sense, only secondarily about dams and reservoirs. The "heart of the story," as Harvey notes, was the "dire threat to the national park system." There was little reason to think that the defeat of Echo Park Dam would be duplicated for proposed dams in areas that were neither national parks nor national monuments.[7]

For very good reason, Denver Water's leaders did not see in the fate of the Echo Park Dam any foreshadowing of the future of their own projects. In the mid-1950s, the scale of the legislative revolution that would take place in the late 1960s and early 1970s was beyond anyone's prediction. Harvey reminds us of what a different world the West of the 1950s was: "It should be emphasized that this controversy took place long before the 'environmental revolution' of the 1960s [and 1970s], and that the opponents of the dam had no legal weapons to block or delay the project." No advocate in the 1950s could deploy the "environmental impact statements or other legal weapons" that would be "available to a later generation of environmentalists."[8]

Echo Park was a very important story, but it was far from an unmistakable demonstration that the conditions of water development had undergone irreversible change. Even to alert observers, the traffic light at the intersection of water development and natural preservation might have turned yellow, but it certainly had not turned red. With the successful and uncontested completion of Gross Dam at the center of their attention, the Denver Water

Board proceeded under the impression that there was no reason to stop, and even a reason to speed up.

♦♦♦

Although it was located on the Front Range, Gross Reservoir played a crucial role in the whole system of diversion from the Western Slope. If the Water Board could not put the water that traveled under the Continental Divide through the Moffat Tunnel to immediate use, they would have to spill it—release it to flow down the South Platte and beyond the reach of the city. Importing water for the benefit of downstream Nebraskans understandably struck the officials of Denver Water (and the representatives of the Western Slope!) as waste. Thus, the capacity to store diverted Western Slope water on the Front Range, keeping it available for Denver's needs, was a necessary and urgent element of the agency's plans. With Denver Water's future-oriented mind-set, constructing Gross Dam was only one step in the larger enterprise of making sure that there would be sufficient storage on the Front Range to avoid "waste" and to hold enough Western Slope water to meet future demand.

Without those places of storage, owning the rights to water was an abstraction with little material meaning. The complex of major dams built on the South Platte River and South Boulder Creek over the course of the twentieth century—Antero, Eleven Mile, Cheesman, and Gross, plus the holding basins of Marston and Ralston—served as the Water Board's principal buckets on the Front Range, collectively storing almost 270,000 acre-feet. The massive Dillon Reservoir west of the Continental Divide held nearly that much again. The flow of the water that came from the Western Slope had to be managed in Front Range reservoirs and strategically released by calculations of its abundance and the intensity of demand.[9]

Once Dillon Reservoir went into operation, the Denver Water Board confronted an unusual problem of overabundance. During the high runoff season of the spring and early summer months, when Denver Water could draw on its direct flow rights to divert a certain flow of water straight from the Blue River into its system without first storing it in Dillon (storage rights for filling the reservoir were applied separately), the agency could deliver more water through the Roberts Tunnel than it could store in its lower South Platte system. What to do? The board sought the remedy in a long-contemplated storage project: a dam at the confluence of the North Fork of the South Platte and the main stem of the South Platte, a site known as Two Forks.[10]

In its topography and geology, Two Forks looks like a model dam site.* A mile and a half below the meeting of two rivers, the canyon pinches in, in a fashion reminiscent of the Cheesman Dam site, creating a narrow gap, nicely sized for a dam, measuring only a hundred feet across at its floor. The floor is free of faults, providing an excellent base for anchoring a dam. George Cranmer, the former city manager of public works who oversaw the construction of the Moffat Tunnel, called it "a perfect damsite" in 1965. Even Daniel Luecke, one of the leading voices opposed to the reservoir, had to agree in 2009 that the spot is "an ideal place to build a dam."[11]

With an initial claim made in 1905 by the Water Department's predecessor, the Denver Union Water Company, the Two Forks site had been under consideration since the early years of the twentieth century. In 1966, just a year after a devastating flood on the South Platte had rampaged through Denver and other metro area communities along the river, the Bureau of Reclamation put forward a plan for a dam at Two Forks. The dam would be built and operated by the bureau with the understanding that Denver (and its suburban neighbor Aurora) would be the primary beneficiaries of its water storage, but that it would also provide flood control and hydroelectric power generation.† In 1970, an organization called the Rocky Mountain Center on Environment (ROMCOE) responded to the studies undertaken jointly by the bureau and the Denver Water Board. "The project," the ROMCOE report asserted, "hasn't been properly treated on an ecological systems basis or on a regional planning unit basis." Imagining the dam's consequences triggered alarm over urban and suburban expansion. "In 30 to 50 years will there be nothing but this transplanted urban hustle and bustle?" ROMCOE asked. "We are contemplating the drastic alteration of the personality and character of hundreds of square miles

* While the site did indeed present the advantage of a narrow canyon, where a dam would back water up into a broad valley, the hydrological efficiency of the project came into dispute. Environmentalists argued cogently that there was a gap between the dam's capacity for storage and the actual firm yield of water that it would provide. The critics made a convincing claim that for a good share of the time, the reservoir would be less than half full. (Personal communication from Daniel Luecke to Jason L. Hanson and Patty Limerick, December 27, 2010.)

† The 1965 flood also spurred Congress to allocate funds for Chatfield Dam, with flood control as its primary purpose, and inspired Denver to create a greenbelt along the river where it runs through the city core in an effort to reduce the potential for damage in any future floods.

. . . is this [dam] another element of the sprawl of the megalopolis which will be smeared from north to south?" ROMCOE did not take an explicit position on Two Forks, but asked for greater consideration of "fish, wildlife, soil erosion, recreation development, [and] land-use planning." Within a few years, another controversial project would push the Two Forks Dam out of the center of attention, but between ROMCOE and the objections expressed by mountain residents, several well-staged rehearsals of the dispute over its construction had already been performed.[12]

In the early 1980s, the intensity of Denver Water's interest in Two Forks picked up. In the 1983–1984 season, the environmental coordinator for the department's planning division, Robert Taylor, lamented that Denver "lost about a million acre-feet of spring runoff because there was nowhere to store it." This state of affairs wore on the patience of managers who wanted to capture every drop they could. A dam at Two Forks could capture much of that water and hold it for use in dry times. Planners found an additional advantage in the fact that the Forest Service and the Denver Water Board owned most of the surrounding land, lessening the prospects for opposition from prior residents.[13]

While its advantages as a dam site struck advocates as unmistakable, Two Forks remains as dam free as Echo Park. Unsubmerged, the place stands as a monument to a major victory for the environmental movement, a crucial exercise of federal support for the preservation of natural settings, and a decisive setback to Denver Water's ambitions. For the better part of a century, the Denver Water Board had worked to orchestrate the resources of several watersheds encompassing roughly thirty-one hundred square miles on both sides of the Continental Divide.[14] With extraordinary determination, its engineers and workers had bored miles of tunnel through the Colorado mountains, impounded hundreds of thousands of acre-feet of water, laid thousands of miles of pipe, dug hundreds of miles of canals, and installed thousands of valves to secure a reliable and abundant water supply for the people of Denver. For all this energetic activity, in the last half of the twentieth century, a new set of obstacles, hurdles, restraints, and regulations appeared in Denver Water's path.

Some of these new elements had figured in the story of the Blue River diversion, but Denver Water had still achieved its primary goals in that episode. In the struggle over Two Forks Dam,

The South Platte River's flood in June 1965 left some dramatic wreckage in its wake, such as this pileup on West Alameda Avenue. It also prompted the US Bureau of Reclamation to put forward a plan for a dam at Two Forks, a site that the Denver Water Board and its predecessors had been interested in since the first years of the century.

Denver Water would finally confront opposition that it could neither overpower nor outmaneuver. Reflecting on the fact that the Two Forks Dam would not be built, the Environmental Defense Fund's Daniel Luecke concluded that "we have at this point seen the end of the big dam era."[15] The shift begun at Echo Park had come to a definitive conclusion at Two Forks.* As a lasting demonstration of the scale and impact of changing American attitudes toward the use of natural resources, the defeat of Two Forks signified beyond doubt that the rules had changed, and the big dam era was indeed over.

♦♦♦

Why, in the 1980s, did the Denver Water Board not try to alter its course and steer around these new obstacles? For an agency operating under the mandate to provide adequate water for the citizens of Denver, *changing its course* could well sound like a synonym for *betraying its mission.* Exemplifying that point of view, Denver Water's chief legal strategist and public voice, Glenn Saunders, greeted the emerging shift in attitudes and policies with neither cheer nor submission. Earlier, in the late 1950s, as Congress began deliberations on a bill for the protection of wilderness areas, Saunders found the whole idea preposterous: "How," he asked, "can any right-thinking American give serious consideration to wilderness legislation which will hamper the water development necessary to unlock [the West's] vast potential?"[16] The federal government, he said on behalf of the board to a Senate committee in 1959, was wavering in its willingness to grant and lend money for reclamation projects. And now federal officials were pressuring the Denver Water Board to change its operations to leave more water in rivers for recreational use and for the needs of aquatic life. The US Fish and Wildlife Service, a part of the Department of the Interior, had demanded that the board maintain a minimum stream flow

* The end of the big dam era was signified by a definitive shift in public attitudes toward major impoundments, not by an instant moratorium on all dam projects. In the decades following the defeat of Two Forks, two additional big dam projects were completed. The Army Corps of Engineers built the Seven Oaks Dam on the Santa Ana River in southern California, a project that was approved by Congress in 1986 and completed in 2002. And the Bureau of Reclamation completed the Ridges Basin Dam in southwestern Colorado in 2007, part of the Animas–La Plata project that was approved by Congress in 1968 to meet US obligations to the Ute Mountain Ute and Southern Ute tribes that date to 1868. There are currently no additional big dam projects under construction in the West.

of thirty-five cubic feet per second in the Williams Fork in order to sustain trout. Such fisheries generated significant income for the local community and the state as a whole. Unmoved by these appeals, Saunders felt that the fish could just figure out a way to get by with *two* cubic feet per second. As always, to Saunders, Denver Water's job was to provide water for the city, a job that left no room for concerns about the Western Slope's fortunes or fish, and that translated the demands of federal agencies into unwelcome nattering.[17]

So Saunders was finding federal intrusions to be an irritation in the 1950s. To use colloquial but emphatic phrasing, he hadn't seen anything yet.

The escalation of environmental regulation and litigation was truly a change of eras for the nation and particularly for the American West. Not long ago, western historians took part in a furious fray over whether the year 1890, long designated as the end of the frontier, actually represented a major watershed, dividing a frontier West from a postfrontier West. Even as professorial feathers and fur flew in this arcane spat over chronology, evidence was accumulating that a far more significant divide in time was taking shape. The passage of the environmental laws of the 1960s and 1970s divided the phases of eestern history in the most dramatic and literally down-to-earth way. Consider the major components of this revolution, which, taken together, amounted to an entirely new set of ground rules: the Wilderness Act of 1964, the Wild and Scenic Rivers Act of 1968, the Clean Air Act of 1970, the National Environmental Policy Act of 1970, the Clean Water Act of 1972, the Endangered Species Act of 1973, and the Federal Land Policy and Management Act of 1976.

State government played its own role in this revolution. In Colorado, the environmental spirit of the times led the legislature to pass HB 1041, a law that gave local governments significant power to regulate "areas and activities of state interest" (and is therefore often referred to as the Areas and Activities of State Interest Act). The act empowers local governments to exercise some authority over the development of resources within their jurisdiction, thus putting a check on the Denver Water Board's ability—which the city agency had enjoyed since the passage of Article XX in 1904—to reach beyond the city borders for water without the approval of the impacted citizens. The Denver Water Board challenged the law in court, claiming that it amounted to an unconstitutional delegation of power to local governments, but in 1989 the Colorado Supreme Court finally and decisively upheld HB 1041.[18]

Whatever case old-school historians tried to make for the transformation wrought by the end of the frontier in 1890, that transformation was dwarfed by the impact of this extraordinary episode of lawmaking in more recent times. And when it comes to choosing the prime example of the game-changing consequences of these laws, the story of the Denver Water Board easily qualifies for the list of the finalists.

Dams, reservoirs, and tunnels through mountains, and the fights and struggles they trigger, have a certain scale and drama. By contrast, a person who would like to designate water treatment plants as gotta-have-it items in the cultural literacy list of the informed citizen will have to work vigorously on her sales pitch. Still, nothing is more important to citizen well-being than these seemingly dreary facilities. Denverites may have happened on to the good fortune of relying on a water supply that comes more or less directly from the pure snows of the Rocky Mountains, with very little of the pollution that rivers pick up as they flow through farmlands, towns, cities, and industrial areas. Denver's situation is, in other words, very different from New Orleans's or Saint Louis's. But even Denver's water requires treatment to make it reliably safe and potable.

Despite earnest and prolonged efforts to build adequate facilities, the Water Board's ability to process water struggled to keep pace with public demand. In the 1970s, growth in the metropolitan area stretched to its limit the department's capacity to treat raw water. Hot July days in 1972 and 1973 pushed water demand beyond the treatment system's recommended capacity of 460 million gallons per day, spiking to a record 506 million gallons on July 6, 1973, as Denverites coped with a high of 103°F. Thirsty customers drew enough water through the pipes during these heat waves to deplete holdings in the reservoirs and raise fears among water managers that a day might come—and soon—when they would not be able to treat enough water to meet peak summer demand.[19]

In its customary manner, the Water Board responded to the threat of shortfalls with a plan to expand the system. The board proposed a $160 million bond issue that would finance the construction of a reservoir and a treatment plant in the foothills southwest of Denver. With the Strontia Springs Dam situated in Waterton Canyon, a reservoir would settle water diverted from the South Platte before it moved to the anticipated Foothills Water Treatment Plant near Roxborough Park south of the metro area. The bond measure was defeated in 1972, in large part because environmentalists saw the proposed 500-million-gallon-per-day plant—which would give the water agency a capacity to treat far more water than it actually had in storage—as a sneaky prelude to building Two Forks Reservoir, which they worried would mean increased diversion from the Western Slope and more sprawling growth around the metro area.[20]

After another scorching summer, the Water Board tried the measure again in 1973, this time with a promise that the initial plant would be built to treat only 125 million gallons per day, with expansion to its full 500-million-gallon capacity coming only as needed in later phases. The Water Board's promises reassured enough voters to win approval for the bond measure in the fall of 1973, but it did not satisfy all of the agency's critics. Even if it started small, critics were quick to observe, the Foothills Plant was not designed to *stay* small. The dam at Strontia Springs and the tunnel conveying water from the reservoir to the plant were going to be full-size from the start, making the later expansion of the plant itself look foreseen and intended. A big question hovered over Foothills: since an increase in Denver Water's capacity to treat water would exceed the current raw water supply, where was the additional quantity of water for the full-size plant going to come from? With the answer to that question left hanging, the Foothills Plant looked more and more like the entering wedge for a much larger project—one that would include bigger diversions from the Western Slope—in the near future.[21]

♦♦♦

In the worldview of Denver Water's leaders, all their foresighted anticipation of future needs should have elicited *not* condemnation, resistance, and opposition but gratitude, support, and applause. In the twenty-first century, it may seem that these leaders confronted an earthquake in public opinion and misread it as a few inconsequential tremors. In truth, the scale of this shift in attitudes toward growth was almost beyond comprehension. For the better part of two centuries, Americans had looked at the American West and found "manless land calling out for landless men." Getting more people into the West—promoting the region, advertising its resources, recruiting settlers to occupy more homes—served as the operating definition of progress. In the middle of the twentieth century, the suggestion that the definition of progress was on the verge of reversing, recasting the formerly welcomed increase in population as a *problem*, was psychedelic (to use the valuable term that would soon emerge from the counterculture). But changing attitudes toward growth, along with the new powers given to federal agencies, brought into being novel alliances and oppositions that challenged the inherited definition of progress.

The extraordinary new environmental laws were the most important—and mystifying—manifestation of this changed world. Given the warm feelings that many members of Congress have for the spotlight, it is a temptation to think of the occasion of passing a law as the conclusion of the main story, with a less important epilogue trailing behind and trudging through the bureaucratic process of putting the law into practice. A more realistic understanding of innovative legislation emphasizes the implementation of the law and reduces the story of the law's passage to a prelude. No one, in other words, knows exactly what a law is going to mean until it has been applied on the ground level, until earnest bureaucrats have tried to exercise the powers seemingly granted to them by the law, and until individuals and groups who disagree with the bureaucrats' interpretation of their mandate and power have talked to reporters, rallied fellow citizens, hired lawyers, and awaited and accepted the decisions of judges. Denver Water's efforts to build the Foothills Treatment Plant and the Two Forks Dam thus offer case studies of vital importance to Americans trying to figure out how the 1960s and 1970s environmental laws had transformed their nation.

As our first step into the labyrinth of implementation, we begin with the recognition that the phrase *federal government* is a term that struggles to hold together a multiplicity of agencies and bureaus that do not always take joy in each other's existence. In order for Denver Water to get permission to build the Strontia

Springs Dam and the Foothills Water Treatment Plant, five federal agencies had to say some version of yes. This meant that every one of those agencies had to reach agreement, not just with the Denver Water Board, but *with each other.*

The first two agencies were pulled into the controversy because the dam and the reservoir would occupy twenty-seven acres of land under the jurisdiction of the US Forest Service (USFS) and twenty-two acres of land administered by the Bureau of Land Management (BLM). With the Echo Park comparison in mind, it is important to note that the Foothills project did not involve land designated as a national park or national monument. On the contrary, both the USFS and the BLM officially operated as multiple-use agencies, so the sanctity of the national park system, the principal issue of Echo Park, would not be raised. For the project to be built, both agencies had to grant right-of-way permits. Issued without controversy in the 1960s, those permits had expired and needed to be reissued. Here is a fine indicator of the sharp difference between two eras: the National Environmental Policy Act had become the law of the land in 1970, and so the reissuing of these permits had, over the passage of just a few years, become a vastly more complicated process, requiring what was sure to be a detailed, closely inspected, and vigorously disputed environmental impact statement (EIS).[22]

At the center of the regulatory transformations taking shape in the 1970s was the Environmental Protection Agency (EPA), created in 1970 by President Richard Nixon (note, again, the timing of the Foothills initiative, with the bond issue passed in 1973 while the EPA was still in institutional infancy). For several years after its origin, the EPA was an idea waiting to be turned into practice and substance. No one had more than an informed guess of how extensive its powers would be, nor how much autonomy its officials would be given to exert those powers in the complicated world of national and regional politics.

When the BLM and the USFS moved toward granting right-of-way permits, the EPA took the occasion to exercise its new powers. Invoking Section 309 of the Clean Air Act of 1970, the EPA brought in another federal organization, the Council on Environmental Quality (CEQ), which is part of the Executive Office of the President, to review the decisions. In making its case to the CEQ, the EPA challenged the validity of the studies of the BLM and the

USFS on grounds that might, at first, seem strangely unconnected to a water project.* The Foothills project, EPA argued, "would make the attainment and maintenance of national ambient air-quality standards in Denver more difficult and perhaps impossible." How, specifically, did the EPA connect a project in storing and treating water directly to damaged air quality? EPA officials found the connection self-evident: "providing water would encourage growth and urban sprawl" and thus recruit more Front Range residents to drive their automobiles hither and thither, befouling the air as they visited, shopped, transported offspring, commuted, worked, and pursued diversion and entertainment.[23]

Another of the new environmental laws brought another federal agency into this scrum, giving the EPA a hold on the permitting process. Section 404 of the Clean Water Act of 1972 required a special permit when the construction of a dam involved inserting fill material into a waterway. The Army Corps of Engineers held authority over the issuing of this Dredge and Fill Permit (commonly, if not creatively, referred to as a 404 Permit by the bureaucratically sophisticated). However, subsection 404(c) gave the EPA the authority to veto any Dredge and Fill Permit issued by the corps (following the pattern, such vetoes are unimaginatively called 404(c) Vetoes). When the corps moved to issue such a permit for Strontia Springs Reservoir, an integral component of the Foothills complex, EPA regional administrator Alan Merson objected, exclaiming that the project's planned capacity was so out of scale with Denver's needs that building it was akin to trying "to kill a flea with an elephant gun." Aligning himself with the environmentalists who perceived Two Forks Reservoir as the unmentioned elephant (certainly not the flea) in the room, Merson threatened to kill the whole Foothills project by issuing a 404(c) Veto of the Dredge and Fill Permit issued by the corps.[24]

Toward the end of the queue of jockeying agencies was the US Fish and Wildlife Service. A much-amended law originally passed in 1934 required federal agencies to consult with the Fish and Wildlife Service on decisions that impacted the fish and wildlife under its purview. This meant that the well-being of aquatic organisms had to be factored into the plans for Foothills, particularly in

* This sentence is a stunning example of the principal sorrow of writing environmental history in the last fifty years. Using the full name of federal agencies is cumbersome, leaving only the alternative of acronyms. An apology for inadvertent literary injury nonetheless seems in order.

terms of the minimum stream flows required to keep those sensitive creatures from going belly-up.[25]*

"When we try to pick out anything by itself," famed preservationist John Muir once said, "we find it hitched to everything else in the Universe."[26] Here, in a rather different sense than Muir had intended, was the central lesson of this new era. To adapt Muir's quotable line to Denver Water's point of view: try to pick out any project by itself and you'll find it connected to a censorious and disapproving federal agency, to an adverse environmental impact, to a chain of consequences that environmentalists will find reprehensible, to big trouble in public relations, and to a tower of paperwork.

Facing an array of federal agencies newly supplied with an expanded and restocked tool chest of regulatory legislation, the Denver Water Board's proposal for a new treatment plant had appeared, depending on your point of view, with the best or the worst possible timing. Here, arriving like a precisely timed train at a station, was a prime test case for defining what the new environmental regime was going to mean in down-to-earth practice. Never before had the board's fortunes (like the fortunes of thousands of other such agencies around the nation) been subject to so much regulatory and legal oversight.

Feistier than many of its counterpart agencies, the Denver Water Board responded to the EPA's veto threats by questioning the authority of federal agencies to determine local land-use priorities. Even if the EPA's premise proved accurate and an increased water supply inherently and inevitably led to undesirable growth—even then—the board asserted that decisions over resource use were properly the responsibility of the local government rather than unelected agency officials. The issue, in other words, had acquired an interesting dimension of contested federalism, as Denver Water undertook to assert the rights of municipalities and states in a world of expanding federal power.[27]

As vexatious as Denver Water was finding its dealings with federal agencies, at least those entities had been called into being by Congress and presidents, placed within more or less defined boundaries, and structured into official hierarchies and chains of command. Considerably more flummoxing was the proliferation

---

* The Federal Energy Regulatory Commission was the sixth agency with a hand in this permitting process, but really, the reader's patience has been sufficiently tried already.

of environmental groups, brought into being by citizen activists and proving to be capable of putting up very effective fights. The president of the Denver Water Board Charles Brannan offered a memorable dismissal of his environmentalist opponents: "Negotiations with people who have no authority in the matter and whose agreement would be utterly unenforceable in any administrative or legal form would be without purpose and be an indefensible waste of public funds." While there is no reason to admire his intransigence, only those who have had their sensitivities surgically deadened can avoid a moment of empathy for Brannan's frustration in deciphering a disorienting new world. He and his colleagues were now encountering peppy groups of people who gathered at coffee houses, living rooms, trailheads, natural food stores, art galleries, bookstores, and various public meeting areas, who formed themselves into teams, committees, and communities, and who then fought Denver Water like banshees. How was an accomplished and well-established utility in its mature years to take seriously these "people who have no authority in the matter"?[28]

One way to deal with them was, of course, to sue their pants off.

The Foothills controversy produced two suits in federal court. The first one, filed by Denver Water in US district court in Denver, began by rejecting the conditions of the BLM permit as "unacceptable and illegal." Since the agency held legal rights to the water that would be stored at Strontia Springs, Denver Water's attorneys asserted, no federal agency could put stipulations on its use or release. The second suit, brought by a coalition of environmental groups in US district court in Washington, DC, claimed that Denver Water's plans were in violation of the Federal Land Policy and Management Act of 1976 (known by the clunky acronym FLPMA), a law that codified an emphasis on conservation goals in the BLM's management, and the National Environmental Policy Act of 1970 (NEPA), a law that specified a process for studying and mitigating environmental impacts.

The stakes of the litigation escalated dramatically on August 22, 1978, when the Denver Water Board amended its complaint by alleging that a coalition of environmental activists and federal officials had illegally conspired to "impede, delay, and prevent" the Foothills project. The board named fourteen specific environmental groups and seventeen individual federal officials as conspirators and asked the judge to assess $36 million in damages against them. At the top of the list were EPA regional administrator Alan Merson and his deputy, Roger Williams, whom the Water Board singled out for $3 million in damages.[29]

At this point, the Foothills squabble acquired its greatest national significance. Environmentalists feared that if the Water Board's complaint was upheld, the precedent would permanently undermine NEPA, allowing large institutions to silence antagonistic environmental groups through the threat of a catastrophic lawsuit.[30] Such a judicial finding, so early in the life of the new environmental law, would have had enormous consequences for the whole matter of public participation in the decisions of federal agencies. By declaring that the environmentalists' opposition to the Foothills Plant rendered individuals and groups punishable and liable for the damage they had allegedly inflicted on the agency, Denver Water had aimed a sharp kick at the First Amendment.

♦♦♦

With the permitting process inching through lawsuits and red tape, Denver Water officials exercised their own free speech, ominously declaring that delay and uncertainty would have grim consequences for Denver's neighboring communities. Reluctant to take on any new service obligations, in June 1977 the board moved to impose lawn-watering restrictions on its customers and cap the number of new residential taps outside the city at only fifty-two hundred per year, meaning that Denver had cut by a third the number of customers to whom it would normally have extended service. This constraint seemed to echo the Blue Line, sending a shock through the metropolitan area's economy.[31]

As homebuilders and suburban officials looked ahead to restricted water services, they pressured the EPA to find a resolution. The environmental groups who had filed the lawsuit in US district court in Washington were short on money, and uncertain of their chance for victory in court. The regional staff of EPA, meanwhile, was feeling out the limits of its authority, uncertain exactly how far its powers extended. While the Clean Water Act did give the young agency the power to block the Foothills project with a 404(c) Veto, all of the interested parties were well aware that the EPA had never yet actually exercised that power. These circumstances, combined with another scorching summer in 1977 that again taxed Denver's water treatment system to its limit, gradually created a receptivity to negotiation. The Denver Water Board was the most resistant to the idea, and this led to a turn of events that at once broke the usual rules of proper mediation and also made the mediation successful.[32]

Theory required that mediators be scrupulously neutral. Stepping forward as the prospective chief negotiator, Colorado congressman Timothy Wirth was already known for his longstanding support of the Foothills project. Rather than derail the negotiation, however, Wirth's unconcealed position proved to be a key asset. The Denver Water Board expected Wirth's inclinations to work to their benefit, thus leading them to drop their resistance and agree to take part in the negotiations. Wirth, moreover, was indefatigable in his preparation, engaging in multiple consultations before he tried to convene any of the antagonists in a larger conversation. And then, in another inspired move, Wirth undertook a first round of negotiation with the three principal contestants: the Environmental Protection Agency, the Army Corps of Engineers, and the Denver Water Board, shepherding those three to an agreement before bringing in the Bureau of Land Management, the Forest Service, the Fish and Wildlife Service, and the environmental activists. While this phasing of participation risked arousing the resistance of the parties left out of the first round, it also made it possible to work through the most difficult conflicts with fewer participants and also to create a momentum that would make it hard for any one official or agency to run the risk of condemnation as the recalcitrant and bad sport who chose contention over problem solving.[33]

In the Foothills Agreement, finalized just after the new year in 1979, the Water Board made important concessions in exchange for the opportunity to build Strontia Springs Dam and the Foothills Water Treatment Plant. The Army Corps of Engineers would issue the necessary Dredge and Fill Permit, and the EPA would not object. The permit would, however, carry requirements for both a new Denver program in water conservation and also for efforts to mitigate the impacts of the project. The board agreed to give more attention to public opinion through the creation of an advisory committee. Both the suit brought by the environmentalists and the suit brought by Denver Water would be ended, and Denver Water would admit "that the environmental defendants have asserted

Completed in 1983, the Strontia Springs Reservoir in Waterton Canyon settles water diverted from the South Platte before conveying it to the Foothills Treatment Plant.

their opposition . . . in good faith and within their Constitutional and statutory rights," and were thus "not proper parties to this litigation." The agency, moreover, would pay for the environmentalists' attorneys' fees. When and if Denver Water proposed another big project, the permitting process would require not just an EIS for that particular project, but a *systemwide* environmental impact statement, assessing all of Denver Water's facilities and structures.[34]

Despite the concessions to the environmentalists, the board had achieved its goal of expanding its system with the Foothills Treatment Plant and the Strontia Springs Dam. In June 1979, they signaled their satisfaction with this outcome by easing the moratorium on new taps, increasing by eighteen hundred the number the agency would install to seven thousand annually. When the plant came online in the summer of 1983, the board eased the watering restrictions that had been in place since 1977.[35] Into their future battle over Two Forks Reservoir, the leaders of Denver Water would carry the lessons they drew from the events of the Foothills campaign, lessons that did not include methods and techniques for the graceful acceptance of defeat. On the contrary, the lessons of Foothills for the Denver Water Board matched the lessons of the Blue River Decrees: an organization facing opposition and frustration could prevail over these afflictions by threatening severe consequences if its demands on behalf of the public were rejected, submitting to negotiation, and in the end making a concession or two. Even if the pursuit of the goal had to follow a circuitous route, the agency had secured permission and built the treatment plant. In 1979, the Denver Water Board had managed to postpone a full reckoning with the changed world of federal authority over local water operations and with the remarkable rise in influence of environmental groups.

♦♦♦

Like many twenty-first-century Americans, the leaders of the Denver Water Board in the last decades of the twentieth century held assumptions, derived as much from stereotype as from evidence, that the stances of the two major political parties toward development and environmental impact were unmistakably distinct and even opposite. Despite the fact that much of the legal framework for the environmental revolution—including the Clean Water Act,

When the Foothills Treatment Plant came online in 1983, it increased Denver Water's capacity to supply water to household taps throughout the metropolitan area. Although its completion marked the end of one controversy for Denver Water, it was soon followed by a much larger controversy over the Two Forks Reservoir.

the Endangered Species Act, and the executive order creating the EPA—bears the signature of Republican president Richard Nixon, the Republican victory in the 1980 presidential election seemed to signify the return of a friendlier federal stance toward resource development. The Denver Water Board saw a good omen in President Ronald Reagan's appointment of two Coloradans to key environmental posts in his administration, naming James Watt as secretary of the interior and Anne Gorsuch as EPA administrator. The board therefore made an approach to authorities in Washington, DC, testing out various ideas for projects.

They had plenty to talk about, with several projects on the drawing board in addition to Two Forks. In 1963, the Water Board had filed for additional water rights in the Williams Fork basin with the plan of diverting the water into an enlarged Moffat collection system. And in 1968 and 1971 they had filed for a suite of new water rights on the Colorado River, Piney River, Eagle River, Gore Creek, and Straight Creek. The Eagle–Piney, Eagle–Colorado, and East Gore projects, as they were individually known, would collectively bring into being an elaborate series of new reservoirs and tunnels on the Western Slope that would convey roughly 250,000 acre-feet of water (about the size of another Dillon Reservoir) under Vail Pass and through Roberts Tunnel into the South Platte, where the water might be stored in the imagined Two Forks Reservoir.

The fact that the US Forest Service, cheered by environmental groups and supported by the courts, was maintaining a stiff resistance to the Williams Fork enlargement might have discouraged water suppliers with less pluck than Denver Water. The fact that Congress, at the behest of Colorado's environmental community, had created the large Eagles Nest Wilderness Area in 1976, smack in the middle of the proposed Eagle–Piney, Eagle–Colorado, and East Gore projects, might have seemed like a deal breaker to less spirited water agencies. In hindsight, it is easy to wonder why the Denver Water Board did not interpret this swell of opposition as a warning shot across the bow from an environmental community carrying growing clout. But interpreting oracles is a tricky business, and when the Denver Water Board reflected on its successes with Foothills and the Blue River, it was just as reasonable to anticipate the agency's eventual triumph through strategy and persistence. Despite daunting challenges, the Denver Water Board pressed its case confidently in Washington and at home.[36]

Looming largest of all of the Denver Water Board's endeavors was Two Forks Reservoir. Colorado's governor Richard Lamm tried to forestall another round of bitter conflict by offering to bring the contestants together for a negotiation. The Denver Metropolitan Water Roundtable assembled a Noah's Ark of thirty stakeholders in the metro area's water future: "county commissioners, suburban mayors, municipal water suppliers, South Platte irrigators, land developers, Colorado River counties and towns, and environmental organizations" were all represented, recalled Daniel Luecke, himself one of the two environmental delegates offered a seat at the table. The representatives of the environmental community had been chosen from a coalition of groups operating together as the Colorado Environmental Caucus,* who took as their mission the task of looking for alternatives to Two Forks "that would provide comparable amounts of water by less environmentally damaging means."[37]

Beginning its meetings in the winter of 1981–82, the Denver Metropolitan Water Roundtable was given the ambitious goal of achieving consensus on the future water supply for the area. Unable to get the environmental representatives to endorse Two Forks, Denver Water tried, energetically but unsuccessfully, to force the environmentalists out of the roundtable. The efforts of the Colorado Environmental Caucus to win support for alternatives to Two Forks, meanwhile, had hit the brick wall of Denver Water's opposition. Satisfactorily stymied by 1983, the roundtable took a break.[38]

With an unimpressive record on achieving consensus, the roundtable had, nonetheless, produced very significant results. The Colorado Environmental Caucus had come into its own with major gains in expertise, strategy, and public visibility. The broad environmental coalition had—much to Denver's chagrin—assembled a staff of credentialed and qualified experts, capable of credibly

---

* A striking range of organizations from the national to the local level came together in this informal collaboration. By the time the controversy over Two Forks had fully run its course, the caucus had counted among its members the National Wildlife Federation, Colorado Mountain Club, Trout Unlimited, Environmental Defense Fund, League of Women Voters, Wilderness Society, Sierra Club, Sierra Legal Defense Fund, American Wilderness Alliance, Friends of the Earth, Colorado Open Space Council, National Audubon Society, Concerned Citizens for the Upper South Platte, Western River Guides Association, Colorado Whitewater Association, Colorado Environmental Coalition, Western Colorado Congress, and more. Daniel Luecke was the head of the Environmental Defense Fund's Rocky Mountain office, and the other delegate to the roundtable, Bob Golten, was an attorney with the National Wildlife Federation.

challenging the Water Board's calculations and projections with sophisticated quantitative methods of their own.[39] Denver did not sit idly by and allow itself to be outflanked by collaboration. The Water Board reached out to old rivals to build its own alliances with the Northern Colorado Water Conservancy District, the Colorado River Water Conservation District, and suburbs throughout the metro area.

The Northern Colorado Water Conservancy District did not oppose the construction of Two Forks, fearing that the alternative was metropolitan municipalities buying up—and drying up—agricultural water rights. And, after extensive negotiations, the Colorado River Water Conservation District traded neutrality toward Two Forks for Denver's support of several Western Slope projects. All three water agencies took the occasion to settle longstanding water rights disputes, including agreements on Denver's Eagle–Piney claims and a consideration of the Western Slope's desire to build a pumpback from Green Mountain Reservoir to Dillon Reservoir, demonstrating that intractable disagreements can turn into valuable negotiating chits when the right moment comes along. But the most consequential alliance that Denver built in support of Two Forks was with the suburbs, which joined together with the Water Board to pursue the construction of the reservoir.[40]

Denver's work to build alliances with its suburban communities led to the Metropolitan Water Development Agreement in 1982. This pact established that the suburban water providers would join with Denver in paying for the mandated systemwide environmental impact statement that, by the terms of the Foothills Agreement, had to occur before Two Forks or any other big project could get under way. The Metropolitan Water Providers, representing this coalition of suburban municipalities and water districts partnering with Denver, would pay 80 percent of the costs and the Water Board would pick up the remaining 20 percent; water from the project would be divided in those same proportions. Only a generation after the Blue Line controversy, the agreement marked an extraordinary moment of collaboration in a relationship historically structured by distrust.[41]

The new alliance between Denver and its suburbs responded to several trends that were reshaping the dynamics of the metro area. All over the nation after World War II, major cities found themselves embedded in rings of suburbs. As Denver grew, the suburbs dramatically outpaced the city in growth. Political scientist Brian Ellison summed up the numbers: "In 1940, 28% of the metropolitan population lived outside Denver. By 1960, that number climbed to 49% of the total metro population; by 1986 it was 75%." This striking shift in proportions was certain to rattle preexisting arrangements and allocations of authority and influence.[42]

For several decades, annexation served as the key process for setting the terms of the relationship between Denver and the growing communities in its vicinity. People in new areas of residential development had two big reasons for pursuing annexation to the city. First, Denver absorbed the debts of the special district governments of those areas; this local debt was diluted in the much larger (and growing) pool of Denver taxpayers. Second, unless a suburban municipality had negotiated favorable terms for a permanent water service agreement after the adoption of the 1959 Denver City Charter, the Denver Water Board charged customers outside of the city limits a higher rate for water; annexation thus made water more affordable and spared the annexed community the burdens of developing its own source of supply. While annexation was a process conducted by the city government, water was an important element of the transaction, and thus, as Brian Ellison remarks, "It was through this process that the Denver Water Board became embroiled in the annexation debate." In the 1950s and early 1960s, in the push to build new, large water-supply projects, Denver Water would "deny suburban areas access to its water." Strategically, "the board then blamed the denial on shortages of water supply, which could be solved by the development of the project under consideration." These episodic moratoria played a significant role in the antiannexation movement by "infuriat[ing] suburban communities and increas[ing] suburbanites' mistrust of Denver."[43]

Water was not, however, the most significant factor shaping the workings of annexation in the Denver metropolitan area. "No discussion of the settlement patterns of the American people," the great historian of suburbia, Kenneth Jackson, declared, "can ignore the overriding significance of race." The desire of twentieth-century white Americans to live in places sequestered from the racial and ethnic diversity of core cities was a driving force for the growth of suburbs. As urban historian Jon Teaford put it, "The political boundaries, as well as social, economic, and ethnic frontiers, divided urban America."[44]

As historian Frederick Watson summarized a complicated story, "When the black population began to expand in the early 1950s, the Denver School Board started segregating black students and teachers in certain schools in the district." Over the next years, the Denver School Board, in Watson's words, "employ[ed] a policy of racial containment and stabilization that was designed to prevent white middle-class flight from the city." In 1969, after various protests against the inferior education provided by the segregated schools, African American parents filed the lawsuit *Keyes v. Denver School District No. 1.* With a decision in June 1973, "the [US] Supreme Court for the first time ordered busing for a city outside the South" and "served notice that henceforth even subtle racial discrimination could be a dangerous game for a northern school district to play."[45]

Desegregation and busing triggered consequential changes for local annexation practices. Traditionally, annexation brought new areas into the Denver Water Board's service area *and* into the school district of the City and County of Denver. As Franklin J. James and Christopher B. Gerboth summed up the point, "Suburban parents' fear of becoming involved in the city's court-ordered school busing program was a potent source of suburban opposition to Denver annexations." Led by a local lobbyist, Freda Poundstone, suburbanites who did not want their children to be subject to court-ordered busing supported a constitutional amendment that made it far more difficult for Denver to annex its neighbors, requiring that "annexations be approved by voters in the county from which the annexation is being made."[46] The passage of the Poundstone Amendment in 1974 effectively ended Denver's ability to annex territory from the surrounding counties.

The Poundstone Amendment also carried the consequence, unintended by its proponents, of excluding unannexed suburban areas from the Denver Water Board's core area of service. This, in the big picture of post–World War II urban history, created quite a paradox. Teaford succinctly sums up the widespread pattern of the time: "During the postwar era, suburbia's gain was the central-city's loss."[47] However, Denver's situation presented a major twist on that plot: with significant holdings in water rights and with a vast infrastructure that would be impossible to duplicate, the city held assets that the suburbs could not match.

Since few suburban communities had the financial ability to build their own water systems, contracts with the board provided one of their limited choices. With the board steering by the charter's required priority of supplying customers within the city, these contracts imposed higher rates on the communities outside the city. And suburbanites recognized that if drought occurred, then their contracts could prove precarious. This context of uncertainty lay behind the Metropolitan Water Development Agreement of 1982 and the unusual experiment with regional cooperation it represented.[48]

As the metropolitan area came together in support of Two Forks, the environmental community remained unified in its efforts to stop the dam. The Colorado Environmental Caucus meticulously identified and appraised the unique natural features of the Two Forks site, made inventories of plant and animal species, and projected what would become of these organisms if a dam submerged the canyon. Ecologists warned that inundation would jeopardize local populations of trout, bighorn sheep, whooping cranes, sandhill cranes, and piping plovers. Peregrine falcons might lose nesting habitat. While the falcons had abandoned local aeries since their initial drop in population from the effects of DDT, they had been making a comeback, and they might repopulate those aeries if a reservoir did not interrupt the process.[49]

Among the species catapulted into celebrity status in the environmental conflicts of the late twentieth century, the Pawnee montane skipper joined the snail darter fish in Tennessee and the spotted owl in the Pacific Northwest as an iconic creature with consequence far beyond its tininess. Measuring less than an inch long and easily mistaken for a moth by the uninitiated, the Pawnee montane skipper exists nowhere else in the world but the Platte Canyon. Because of its very limited range, its advocates thought it a good candidate for listing as an endangered species, a status that would bring into force the habitat protection mandated by the Endangered Species Act of 1973. In a recognition of the political power wielded by these modest creatures, the *Rocky Mountain News* in 1985 ran the headline: "The Pawnee Montana Skipper: Is Its Future Worth a Dam?"* But, as in most other controversies involving a particularly noted and discussed species, this butterfly stood for a vast biological community that would be affected by the dam, including blue grama grass, musk thistle, and other indigenous plants for which

* For a while, the Pawnee montane skipper was so new that no one seemed quite sure what its proper name was, and the Montana spelling was not uncommon in newspaper reports.

the Pawnee montane skipper had a soft spot. The circumstances of creatures far from the site of the dam also had to be taken into account: sandhill cranes and piping plovers that relied on adequate flows downstream on the Platte, and the humpback chub and other endangered fish in the Colorado River that swam in increasingly shallow waters with each diversion from the Western Slope.[50]

♦♦♦

Just as in the Foothills controversy, the issuing of a Dredge and Fill Permit under section 404 of the Clean Water Act became the fulcrum of the conflict. Once again, the Army Corps of Engineers was the key agency in this permitting process. Following on the Foothills Agreement's requirement of a system-wide, not just site-specific, environmental impact statement (EIS), the corps had begun work on such a study in 1981. By 1984, the Denver Water Board was getting impatient with the long, system-wide review process and notified the corps of its intent to file for the necessary permits to go ahead with Two Forks, a move that would force the corps to begin a site-specific EIS. After a round of negotiations, the corps agreed to prepare a joint EIS that would include system-wide and site-specific studies if Denver would delay submitting the permit applications and contribute funds to offset the additional costs of the expanded study. This agreement held Denver at bay until the spring of 1986, when the Water Board went ahead and filed an application to build Two Forks, several months before the corps was prepared to release its study.[51]

When the corps finally submitted its draft of the EIS in December 1986, it served notice that it intended to support Denver's application for Two Forks by issuing the 404 Permit for the dam. In April 1987, EPA officials volleyed back, declaring the draft EIS to be riddled with environmental and technical shortcomings. After more jostling between and among federal agencies, the Denver Water Board and its suburban collaborators, and environmental groups, the corps issued its final EIS for review in March 1988. The recommendation had not changed; the corps believed that the project's environmental impacts could be satisfactorily mitigated and that Two Forks should be built.

By the time this final EIS was submitted for public review, the study's size, cost, and production had taken on epic proportions,

For more than a century, the narrow canyon where the North Fork of the South Platte River converges with the main stream has been thought of as both a spectacular setting and an ideal dam site.

weighing upon the earth as a portly twelve volumes. The technical documentation alone stacked more than six feet high. A Denver Water Board spokesperson explained how the document acquired such bulk: "We did one study and that study called for studying something else and that called for another study until it became just a monster." Participating in this data scavenger hunt were "helicopter pilots, tree counters, butterfly catchers, lawyers, accountants, public relations people, newspaper clipping services, biologists, geologists, botanists, engineers, economists, and even dam builders." The price tag for the study totaled $43 million, a cost that, by the terms of the Metropolitan Water Development Agreement, was split 80/20 between the coalition of Metropolitan Water Providers and the Denver Water Board. The study examined the organisms of the Platte Canyon and weighed the relative merits of dam types. It evaluated wetlands disturbance and erosion acceleration, channel stability below the dam, vegetation encroachment, overbank flooding, sediment transport, recreational opportunities, and the impact of the dam and reservoir on the local economy. Described by Ellison as the "largest and most costly environmental-impact statement ever," the corps' study of Two Forks proved to be the behemoth of a new, and consequential, genre of nature writing.[52]

The National Environmental Policy Act had created a giant, nationwide enterprise in the production of EISs, creating a robust new market for the work of specialists, experts, and consultants. If it had itself been subject to EIS, the flood of paperwork generated by these reports would surely have been appraised as the equivalent, in wasting paper, of a thousand ticker-tape parades.

Denver Water's leaders still had reason to think that the lesson of the Blue River Decrees and the Foothills Agreement—the vindication of persistence exercised through a season of great vexation—still applied. Despite the burdensome process of satisfying federal permitting requirements, as late as 1989 the campaign for the Two Forks Dam did not seem lost. On the contrary, in January, the Army Corps of Engineers announced that it was going to issue the section 404 Dredge and Fill Permit, with a formal motion of intent registered on March 15, 1989. The administrator of the EPA's regional office in Denver, James Scherer, tilted toward recommending approval of the 404 Permit, even though some of the local EPA staff members who had worked most directly on the project disagreed with him.[53]

By a widely held stereotype, bureaucracies are such homogenizing, conformity-enforcing forms of social organization that they leave no room for individuals to affect the course of the lumbering machinery of an agency. As this split between the EPA local staff and their ostensible boss begins to indicate, the histories of the Foothills Treatment Plant and Two Forks Dam offer a compelling challenge to that stereotype. The qualities and convictions of the individuals holding office in the EPA constitute very big factors in both stories. In the 1970s, regional administrator Alan Merson was a thorn in the side for the Denver Water Board. In the 1980s, regional administrator James Scherer was quite the opposite, constituting a burr under the saddle for the environmentalists. But most consequential of all was the distinctive character of Scherer's boss, the national EPA administrator William Reilly, appointed to office on February 8, 1989, by President George H. W. Bush, just as the Two Forks controversy reached its moments of decision.

If the EPA had been unsure of the practicable extent of its authority during the Foothills controversy, by now the agency was confident and assured in exercising its power. When Merson stated his objections to the Foothills Treatment Plant in 1978, the EPA had never actually killed a project with a 404(c) Veto. But by the time Reilly took the helm in 1989, the agency was on a streak, having vetoed eight projects in the previous eight years.[54]

On March 24, 1989, Reilly announced that the EPA would review the corps' decision to issue the 404 Permit, invoking a provision in section 404(c) of the Clean Water Act that granted the EPA veto power over projects that would result in an "unacceptable adverse effect on municipal water supplies, shell-fish beds, fishery areas, wildlife, or recreational areas." When Scherer expressed qualms about backtracking on his support for the project, Reilly stepped around the authority of his Denver-based regional administrator and appointed the regional administrator in Atlanta, Lee DeHihns, to oversee the review process. If it seemed like he was stacking the deck against Two Forks, he was. Reilly's prepared statement about the review made the likely outcome unambiguous: "The proposed project contemplates the destruction of an outstanding natural resource," he explained, declaring that "I do not believe this project meets the guidelines . . . of the Clean Water Act."[55]

DeHihns considered the issue (and, one might suppose, the views of his boss) throughout the summer before publicly announcing his findings on August 29, 1989. Citing the "extremely high fish, wildlife, and recreational value" of the area proposed for inundation, he reviewed a litany of threats Two Forks posed to these assets. His considerations included the South Platte's gold medal trout fishery, threatened and endangered fish in the Colorado River Basin, local aeries of bald eagles and peregrine falcons, elk and mule deer herds in the area, recreational whitewater on the Western Slope, water quality in the Blue and Williams Fork Rivers, numerous species that relied on the river habitat in Nebraska, and the vulnerable Pawnee montane skipper. Of the rare butterfly, DeHihns noted that it could well be "lost before the mitigation methods can be proven to work." Weighing against these damages was the Denver metropolitan area's need for water, but DeHihns accepted the arguments put forward by the Colorado Environmental Caucus and others that practicable (even superior) alternatives to Two Forks existed for providing this water. His considerations led to one clear conclusion: "the adverse effects of Two Forks dam and reservoir would be unacceptable. Moreover, it appears these impacts are partly or entirely unnecessary and avoidable."[56]

Denver and its allies vigorously protested and disputed DeHihns's report and EPA's entire 404(c) review process. Colorado's congressional delegation even asked the Government Accounting Office to investigate EPA's handling of the review. But nothing shook DeHihns and Reilly from their initial conclusion. After several rounds of public comment on DeHihns's findings, administrator Reilly accepted the recommendation and officially delivered his veto of the project on November 23, 1990, making it the eleventh project vetoed by the agency under the authority granted to it in section 404(c) of the Clean Water Act.[57]

In an extensive interview after he left office, Reilly said that his veto of Two Forks was "one of the most controversial decisions I had made." Controversy did not lead Reilly to regret or to ambivalence over his action. When he was asked to name his most significant accomplishments, he responded, "first, the elevation of ecology and the signaling of the end to expensive and wasteful water development in the West, which I think is the message of the veto of Two Forks Dam."[58]

Reilly's veto of Two Forks, along with the backing he received from President Bush, offers another direct challenge to the conventional wisdom that Republicans are inherently supporters of development and opponents of environmental preservation. His earlier career had many elements that gave him particular qualifications for appraising the Two Forks proposal. As a young attorney, he worked for the CEQ, where he helped draft the regulations that would "implement the National Environmental Policy Act and the Environmental Impact Statement procedures." He chaired a task force for the CEQ that in 1973 produced a report entitled *The Use of Land: A Citizen's Policy Guide to Urban Growth*, a text that had unmistakable relevance for Denver. Reilly became president of the Conservation Foundation, which then merged with the World Wildlife Fund (WWF). He was president of the WWF when George H. W. Bush, who had punctuated his 1988 campaign with repeated declarations of his hope to be "the environmental president," asked him to become EPA administrator. Reilly recalled that President Bush "expressed his own philosophy" by saying, "I'm not a rape-and-ruin developer and I'm not for locking everything up, either." When the president told him "that he believed in balance," the new EPA administrator was in hearty agreement: "That's really my philosophy. It's one of integration and reconciliation of priorities." Reilly had not been a career politician, and he characterized himself as an outsider in the Bush administration. He did not come to office with a preexisting constituency. "My constituency," he said, "had to be the country."[59]

Reilly was a man of independence and self-confidence, but he did not, in making this move, have to defy the president who had appointed him. Nebraska's Republican governor Kay Orr had conveyed reports on the swing of public opinion against Two Forks to the president (since the Platte flows through Nebraska, upstream storage would affect the state). And, as a frequent vacationer in Colorado, former president Gerald Ford was well-informed on the controversy and wrote President Bush personally to urge him to review the project critically. One leader of the opposition to Two Forks offered a significant tribute to President Bush: "What this means from Bush is 'I really meant it when I said I was an environmentalist.'" A later commissioner of the Bureau of Reclamation made the point even more emphatically: "It was Bush that killed the Two Forks Dam."[60]

Officials at the Denver Water Board at first saw the EPA ruling as just a setback in a longer engagement—a significant move, certainly, but not conclusive. In their recent experience, legal and regulatory obstacles had *not* been checkmates, but only challenging moments in the game, calling for a clever and enterprising countermove. In earlier struggles, most recently in the Foothills controversy, Denver Water had found expressions of alarm over shortages in the future, with attached threats of service limitations, to be effective. In April 1988, back when the Army Corps of Engineers was preparing to release its final draft EIS, Denver Water had returned to its familiar repertoire, warning of water restrictions, increased reliance on nonrenewable groundwater, and metropolitan discord if the permits to build Two Forks were not issued quickly. In June, learning of the opposition to the project held by local EPA staffers, the Water Board had sent a letter to Governor Roy Romer setting forward a grim scenario for greater Denver's future without Two Forks: "Our projection shows that water now available to the Board will be needed for build out of Denver and no water will be available to provide water service outside Denver."[61]

Based on previous experience, Denver Water officials had reasons to expect that these warnings would spur federal officials to reconsider or inspire politicians to intercede, as congressman Tim Wirth had done with the Foothills project. But this time, Denver Water's alarmist statements simply made it harder to find any avenues for compromise or negotiation. Grim predictions, moreover, left Denver Water's claims vulnerable to refutation by credible experts. The days were winding down when a projection of future need could soar above skepticism and float high above the reach of doubt. The old rhetorical playbook that had once worked so well was proving ineffective in a time when environmental groups had built impressive teams of scientific and legal experts who could challenge Denver Water's data and analyses. Sophisticated and adept at quantitative analysis, opponents of Two Forks could argue that Denver Water's predictions were overwrought and unconvincing.[62]

It was in the arena of public attitudes that Denver Water seemed to be most dramatically losing its grip. Despite their limited financial resources, the environmental groups were far ahead of the agency in drawing on the sophisticated approaches of public relations firms. Before Trout Unlimited enlisted the help of the public relations firm Russell, Karsh & Hagan, polls in 1987 showed 50 percent of local people favored building Two Forks, with only 31 percent opposed. By 1990, the percentages had shifted to 32 percent in favor (a drop of 18 points) and 42 percent opposed. The dam's advocates spent more than $40 million in pursuit of their goal. Trout Unlimited, by contrast, ended up raising $42,000 for the effective services of Russell, Karsh & Hagan. In the updated version of David and Goliath, rather than pick up a pebble and launch it at the giant's head, David took the far more effective approach and hired a premier PR firm and followed its advice.

The environmentalists' campaign, called Why Two Forks?, had the central goal of "creating doubt among the public on the major issues associated with the dam" by raising targeted questions in a cohesive and coordinated media and lobbying campaign. A key component of the campaign involved pointing out and addressing many groups who would be losers if Two Forks were built, including "Denver residents who don't need the water" but who would still have to assume a fiscal burden to secure water for the suburbs. Members of Trout Unlimited took part in radio talk shows, wrote letters to the editor, spoke at public hearings at various stages of the preparation of the EISs, and talked diplomatically and discreetly with politicians and businesspeople. Even though the public relations agency wrote frankly, in a private memo, that this was "guerrilla warfare in public opinion," the outright use of such sharp phrasing was entirely uncharacteristic of the approach. As the agency's memo put it, "Various strategies were considered and rejected as being too inflammatory, too radical, [and] not sufficiently broad based." Throughout the campaign, "there were no name calling media stunts, mud slinging, or wild allegations made." This approach took the wind out of the sails of attempts to discredit the project's opponents as wild-eyed and extreme dissidents who could not be taken seriously.[63]

The idea of alternatives proved to be a particularly valuable arrow in the quiver of the dam's opponents. In fact, the National Environmental Policy Act required a thorough consideration of alternative projects that might supply an equal amount of water or even reduce the need for more water. But the adept and clever way in which the leading environmentalists put this requirement to work in a persuasive manner presents a telling demonstration of an underutilized strategy in environmental advocacy. When Daniel Luecke, the director of the Rocky Mountain office of the Environmental Defense Fund and one of the two Colorado Environmental Caucus delegates to the Denver Metropolitan Water Roundtable, made a statement in opposition to the Two Forks project in 1982, he used the concept of alternatives to make a positive statement rather than a negative one. The newspaper quoted him as saying that his group "favors building a reservoir on the North Fork of the South Platte River between Bailey and Grant." A reader steering by stereotypes might have read that statement as a failure in reporting or even as typographical error. Surely, if Luecke represented an environmental organization, the verb would have been *opposes* rather than *favors* building a reservoir? The approach of accenting alternatives permitted a very purposeful and very effective scrambling of stereotypes and expectations. As Luecke said three decades later, "We decided at the very beginning that . . . we would talk" to reporters and the public "in terms of options and alternatives." This permitted the opponents of Two Forks to shed the image of obstructionists; as Luecke put it, "we were going to present ourselves as problem-solvers." Not every member of the Colorado Environmental Caucus instantly caught on to the wisdom of this strategy. "I had a selling job within the environmental community," Luecke remembered, saying, "Look, don't fight with them about water. You create enemies you don't need. You don't fight with them about water and growth. Tell them we understand [their] problem. We'll find you a solution that we can live with too."[64]

In 1980, the Environmental Defense Fund presented an early statement on alternatives to Two Forks, "Water in Denver: An Analysis of the Alternatives." By April 1988, when the Colorado Environmental Caucus submitted an impressive report, "Metropolitan Denver's Future Water Supply," making a case for alternatives had reached the status of rhetorical art. The subtitle of the caucus's statement embodied the fruition of this strategy: "The Colorado Environmental Caucus's Environmentally and Fiscally Sound Alternative to the Denver Water Board's Application . . . to Build Two Forks Dam." Rather than leaving any room for the report to be condemned as contrary nay-saying, the authors fully acknowledged the need to supply water to the metropolitan region. They rested their arguments against Two Forks, instead, on the obligation to pursue "sound financial planning" in order to relieve the residents of Denver of "undue financial burden." "Two Forks would require enormous up-front capital costs," the report said in its most forceful argument. "In the event that the demand for water in the metro area falls short of the projected growth estimates now being used by Two Forks proponents, the big dam has the potential to become a serious financial liability to the metro water providers and to the communities they serve."[65]

In its first phase of alternatives, aimed at getting a sufficient water supply for Denver through 2010, the caucus proposed seven components. Most of them involved swapping and purchasing water from other organizations like the Northern Colorado Water Conservancy District, while by far the largest component was conservation: making sure all homes had water meters (and thus had to pay for the water they actually used), encouraging low-water landscaping, and systematically detecting and correcting leaks. "While individually these projects would yield smaller amounts of water than would Two Forks," the caucus summed up phase one, "collectively they yield nearly 30 percent more water than would Two Forks at less than half the cost." For phase two, aimed at meeting Denver's demand in 2035, the caucus proposed five bigger alternatives, including the noteworthy suggestion to "enlarge Gross Reservoir" as a means of enhancing Denver Water's capacity to store diverted water on the Front Range.[66]

While it is unlikely that "Metropolitan Denver's Future Water Supply" will become a classic of western American literature studied in English classes, it is nonetheless well worth contemplating as a literary text. It has an unbroken tone of calm reason. It provides no opportunity for the proponents of the dam to characterize the environmentalists as impractical lovers of nature unable to think in serious fiscal terms. It concludes with the pleasant assertion that the fight had arisen from an entirely unnecessary pitting of two compatible ideas against each other, since "water development and proper stewardship of Colorado's natural amenities do not have to be mutually exclusive." Consider the concluding paragraph's adept acquisition of the high ground: "The essential difference between the Denver Water Board's choice of Two Forks and the Environmental Caucus's proposal is one of rigidity versus flexibility. The uncertainty in estimating future needs can be accommodated only by a flexible approach to water supply planning." In public communications, the performance styles of the Denver Water Board and the Colorado Environmental Caucus suggest a grizzly bear trying to catch a hummingbird, with the theoretically much more powerful creature increasingly puzzled as to why, even with its most earnest efforts, it could not prevail.[67]

♦♦♦

The Colorado Environmental Caucus emerged from this contest with a scar or two but generally in prime condition and fighting trim. By contrast, the other major coalition, the Metropolitan Water Providers, was in tatters. This coalition between city and suburbs had offered the possibility of creating a regional approach to the water issues of the entire metropolitan area by overcoming the long-standing distinction between water customers inside and outside the City of Denver. With the defeat of Two Forks and not a drop of water to show for their $40-million-plus investment in the project, the suburban water interests in 1991 filed an unsuccessful suit in federal district court in Colorado challenging EPA's veto. Denver refused to join in the suit. The prospects for metropolitan cooperation, with suburbs and city united in common cause, were in deep disarray.[68]

Explicit and official regional cooperation, bringing city and suburbs into a collaborative enterprise, reached its peak—and then rapidly reached its nadir—in the campaign for Two Forks. Is regret or relief the proper response to that outcome? The proposition that water could be more wisely managed if everyone would stop feuding and participate in a collaborative, overarching structure of governance has some power to persuade. Seeming to represent human nature at its best, cooperation in pursuit of common interests holds a more impressive moral stature and a much more positive image than contention and competition. When a cluster of communities positioned in proximity try to work together, rather than at cross-purposes, the observer's first instinct is to applaud. In water management, fragmentation of authority seems to be an invitation to inefficiency, as different organizations duplicate each other's efforts and forswear the savings of economies of scale.

And yet history does not let us go far in romanticizing the charm and virtue of centralized authority. From the perspective of the Western Slope or of antigrowth activists on the Front Range, a consolidated metropolitan water authority uniting Denver and its suburbs into a common enterprise could have been a terror, a juggernaut of centralized power, the realization at last of the long-running fear of an empire running roughshod over other areas of the state with its consolidated urban and suburban power. Humanity has, so far, made minimal progress in finding the equation to calculate what size or scale of institutional power is most likely to produce well-being and wisdom. Thus, the disintegration of the coalition that pushed for the Two Forks Dam presented a classic mixture of good news and bad news, with the proportions between relief and regret shifting like a kaleidoscope according to whether the interpreter was a Western Slope farmer, a Front Range environmentalist, a suburban real estate developer, or a member of the Denver Water Board.

Perhaps even more to the point, we should not overstate the health and harmony of the Two Forks coalition. When participants in the contests of the 1980s reminisce, they will sometimes remark on the internal tensions and frictions at work inside the city/suburb coalition. Some even offer an intriguing line of contrafactual speculation: if the EPA had given the go-ahead to the Two Forks Dam, the enterprise of actually building the darned thing would have taken a frayed and tattered coalition past the point of repair. In a number of episodes in Denver Water's history, the project of building a structure of laws, court rulings, agreements, terms of cooperation, and assigned powers has made the process of building an actual, literal dam seem by far the easier project. Joining together in support of a water project could not, after all, erase underlying and preexisting tensions. In the years before the Two Forks initiative, Franklin J. James and Christopher B. Gerboth remind us, "Suburbanites decided that remaining separate from the city would permit them to maintain racially and economically segregated communities and schools, and to thereby evade the social and economic problems of the central city." The effort to build common cause and harmony could not overpower a fractured history. Before the EPA decision, the Two Forks city/suburb coalition was showing significant structural flaws of its own, and its commitments might well have lacked the sturdiness, resilience, and durability needed for its members to sit as a united body on the stage at a dam dedication ceremony.[69]

Rather than trying to make a clear determination of the desirability of concentrated—or dispersed—authority, a more fruitful line of thought leads to the conclusion that water management has meant, and will continue to mean, a constant jockeying and repositioning of groups and communities in relationship to each other. There has been no end to the emergence of political demands and economic pressures on the institutions that, like Denver

Water, came out ahead in the race to claim water rights. By the last decades of the twentieth century, a whole new cast of characters had entered the fray: Indian tribes asserting the rights guaranteed to them under the Winters Doctrine of 1908; recreational water users, ranging from operators of rafting companies to managers of urban water parks; sports fishermen and fisherwomen; staffers from state and federal fish and wildlife agencies seeking to protect the habitats of endangered species; National Park Service, Forest Service, and Bureau of Land Management employees concerned about the water supply, especially to support aquatic wildlife, in the territory under their management; and varieties of energy companies with plans, from solar energy to oil shale, that would require water for cooling or processing. The existing legal and governance structures have been put through a vigorous dance in adapting to this sharp increase in the cast of characters gazing longingly at western rivers and streams. That dance shows no signs of slowing down.

And yet the distinct identities and opposed nature of the participants in this dance may become harder to track. "Research has questioned the validity of the city-suburb distinction," report urban historians Ruth McManus and Philip J. Ethington. Older suburbs, and even more recent suburbs, "have been swallowed up by the greater metropolitan development," becoming (Reader advisory: Academic Jargon Episode approaching) "part of a variegated polycentric metropolitan complex." With this trend, "it may no longer make sense to look at urbanization as divided between the kind that takes place centrally and that which is peripheral." Or, as historian Martin Melosi put it, in the nation's emerging metropolitan areas, "the well-accepted and long-standing division between core city and suburban ring was more ambiguous (and obsolete)" than it had been in earlier landscapes. If it should prove possible to rearrange the current fragmentation of governmental authority to reflect this blurring of distinctions in categories and in space, then it might turn out that the Metropolitan Water Providers coalition will be noted as a significant precedent for a historic shift (though only the very cavalier or the very wealthy should consider betting on this).[70]

⸫

With the direct path to metropolitan cooperation closed off in 1990, Denver Water retrenched. To reassure clients *within* their service area, the board members felt that they had to be clear and direct in telling the public to give up on the idea of a future in which Denver Water would serve the entire metropolitan area. In 1992, Water Board president Hubert Farbes declared that "present-day events now compel a new path for Denver and its citizens in management and protection of its valuable water supply system." From its origins, "the Denver Water Department has been the central focus for water development to benefit the entire Denver metropolitan area." This is where the new path veered from the old path: "Denver will not, as in the past, lead the metro area in competing for, acquiring, and developing major new water rights for the entire area." The agency would, however, "continue to serve all present and future suburban users within a fixed 'distributor' service area agreed upon with the Department." But Denver Water would not extend service to new areas.[71]

Following the customs of the oracles of our times, Farbes and other spokespeople for Denver Water now charted their course by consulting recent history. "In the last 25 years," Farbes began his presentation of the new path, "many changes have occurred which alter the assumptions and understanding upon which Denver's water system was built." He then catalogued a series of events and trends: the passage of the Poundstone Amendment and the restriction on annexation; the construction of dams at the best sites for water storage, leaving only sites that would be much more expensive to develop; the transformation of "public values and opinions on water development"; the new environmental laws; the escalating costs of water development; the proliferation of suburbs that could not be annexed; and the "layers of institutional and political complexity" that now characterized the Denver area. All this, Farbes argued, had made it necessary to "adapt our goals and strategies to the new circumstances that face us."[72]

The importance of historical context was so visible and evident that the discourse sometimes had the ring of an undergraduate history course or even a graduate seminar. Colorado's governor Roy Romer demonstrated this rise in fortune of historical thought in early reflections on the Two Forks veto: "Our system for planning and developing water was developed a century ago," he declared. "The public values in Colorado had changed dramatically since the system was established." The core beliefs driving the Two Forks proposal, Environmental Defense Fund hydrologist Daniel

Luecke noted at the same time, "was based largely on nineteenth century ideas that were painfully out of touch with twenty-first century environmental values." Jim Ogilvie, who served as the Denver Water Board manager from 1970 to 1979, offered an aphorism that summed up the widely diffused awareness of history: "The biggest problem with Two Forks is that it was not built soon enough."[73]

People who might have spent their younger, more innocent days thinking that studying history meant only memorizing facts and dates now came to an intense recognition that historical change, in the most concrete way, had escaped from confinement in textbooks to shape the context of their lives. No one could miss the fact that Two Forks had come up for appraisal in an era when big dams, unconstrained growth, and technocratic rescues had lost credibility and persuasiveness, and in a new era, old habits were headed for the sunset.

In truth, the contrast between the era of Two Forks and the era of Cheesman Dam was breathtaking. The absence of regulations, the relatively lean and trim state of the federal bureaucracy, and the public's unmixed enthusiasm for a project like Cheesman in the early twentieth century seemed to be features of life on another planet. The very idea of an EIS, much less its bulk and complexity, would never have figured in the wildest daydreams of the men who planned and built Cheesman Dam.

This transformation had been driven by the invention and adoption of new values more than by the invention and adoption of new technologies or new legal interpretations, a situation that had clearly tested and stressed the cognitive world of Denver Water's engineers and lawyers. When Lee DeHihns, the EPA official in Atlanta, reached the conclusion that Denver could not sacrifice the natural value of the Two Forks site, his judgment drew as much on qualitative analysis and sentiment as it did on data and quantified research. Handed a gigantic EIS to appraise, DeHihns had actually been assigned the far more difficult task of weighing and choosing among the shifting values that human beings assign to the natural environment. No mapping of the Pawnee montane skipper's range, no attempt at calculating future population and projecting metropolitan water demand, no analysis of geological structures, and no charting of the factors shaping the reproductive performance of bighorn sheep could generate a mathematical formula that would supply DeHihns with a certain and indisputable conclusion. However carefully some proponents and opponents tried to present their positions as matters of data and logic, the contest was a struggle over cultural belief and preference.

♦♦♦

"Is there any other point to which you wish to draw my attention?" Inspector Gregory once asked Sherlock Holmes.

"To the curious incident of the dog in the night-time," Holmes responded.

Gregory did not instantly decipher the clue. "The dog did nothing in the night-time," he said, to which Holmes famously responded: "That was the curious incident."[74]

Like that legendarily quiescent dog, dams that did not get built usually do not capture the attention of those of us who lag behind Holmes in acuity. Indeed, the site that would have been inundated by Two Forks Dam is not very articulate when it comes to proclaiming its own meaning and significance. It is a beautiful area, but it is by no definition pristine. A railroad ran through the area in the 1880s, gold prospectors rushed to the West Creek Mining District in the 1890s, and loggers worked all over the woods, filling the river valley with "the whirl and whine of sawmills along the river," in the words of a local historian. When the sound of the sawmills died down, proximity to Denver finally gave the area a lasting, if also far from dynamic, commercial base in tourism. In the mid-1890s, Stephen D. Decker, a small-scale entrepreneur who had been stymied in his search for gold, created a proportionately small-scale resort, capitalizing on the mineral water springs of the area. The name of Decker's Mineral Springs and Resort diffused into the world, until the whole area took on the collective name of Deckers. Other tiny resort settlements sprang up in the area, and an automobile stage company went into operation in 1912, bringing Denverites to enjoy what the local newspaper called the "magnificent playground nature has here provided."[75]

From the 1970s on, Denver Water bought land in the area when it came up for sale, with the agency thus developing an uncharacteristic new sideline in cabin remodeling and restoration. In 1981, the Denver Water Board purchased the original Decker's Resort for $592,798, a transaction weighted with symbolism. The principal water supplier in the Denver metropolitan area was also now the owner and custodian of several pleasant resort properties. Here was a microcosm of a large-scale trend: over the course of the twentieth century, the urban desire for recreation ran headlong into the urban desire for water. Enduring this internal collision of desire, a city would be hard put to operate as an imperial monolith.[76]

The double role of the federal government in this collision represented a contradiction of even greater scale. With support for highway construction and subsidies for home mortgages, the federal government set up the conditions for suburban growth. With Cold War spending that brought into being military bases and defense industries, the federal government set off an enormous ripple effect, boosting the western economy and drawing in a flood of jobseekers. And, by locating offices for many agencies in the Front Range area, the federal government gave the Denver area one of the largest concentrations of federal employees outside Washington, DC.

Thus, when the Denver Water Board decided that they had to expand their system, they were responding to a situation made possible by the expanding powers of the federal government. In short order, the water agency found its plans stymied by the massive increase in the powers of the federal government in environmental regulation. Promoting population growth and then calling a halt to the old customs to which the agency had always turned for solutions, the federal government had Denver Water pinned in a very tight paradox.

Where the North Fork of the South Platte converges with the river's main stem, the most powerful trends of the twentieth-century West also converged, giving the name Two Forks a novelistic resonance. "With the splendid attractions in the way of the wonderful spring water, the scenery and the fishing," the *Mountain Echo* wrote in 1898, "Decker's ought surely to become a very famous place." The right of Deckers to claim the status of a very famous place holds fresh meaning in the twenty-first century. On the old principle of "location, location, location," if the descendants of the priestesses and priests who once held sway at Delphi were to seek a place to set up shop in the twenty-first century, they might think about taking a look at Two Forks.[77]

Climate Change and the Stressful Life of Water Managers

As the world proceeds to get hotter,
The power to predict will soon totter.
The baseline's been battered;
The norm has been shattered.
But everyone *still* wants their water.

CHAPTER SEVEN

Chipping Away at Tradition

The Riddle of Change and Continuity at Denver Water

[In the mid-1970s], Chips hadn't had a great deal of experience being a bureaucrat, but he found he had a flair for it.

—Memorial program for Hamlet "Chips" Barry, 1944–2010

We live in a time of great masculine ambiguity, when we light our fireplaces with a light switch, wax our backs, pluck our uni-brows, and use terms like masculine ambiguity. *It's a time when men search their souls with questions like Does having a man's time of the month make me any less of a man?*

Well, does it?

When that time of the month rolls around and you find yourself setting your automatic lawn sprinklers lower—because it's spring, not summer, do you feel ashamed or do you feel proud? Do you wonder if the kind of men who walked on the moon would still call you brother? Or do you say to yourself, "I am using only what I need, just like my father and his father before him"? Then throw your head back and bellow: "I have a man's time of the month and I don't care who knows it." If you can do that you are still a man, my friend. Very much a man.

A message from Denver Water, reminding you to adjust your sprinklers and use only what you need.

—Public service radio announcement delivered in a deep and resonant voice, with stirring musical accompaniment, 2010

"Thirty-five feet down, on the bottom of a concrete tank filled with a million gallons of bitterly cold water, lay a body." Thus begins a recent book on current troubles in the nation's and world's water supply, Alex Prud'homme's *The Ripple Effect: The Fate of Freshwater in the Twenty-First Century.*[1] Recognizing that the topic of water management will not instantly grab every reader, ambitious authors are understandably tempted to leap into the pool with a first sentence freighted with murder and mayhem, a tactic inspired, it seems likely, by the lasting success of the movie *Chinatown.*

No one ever figured out who killed the hydrochemist who worked in a water purification plant in New Jersey, the individual whose sad ending opens *Ripple Effect.* Although it was impossible to draw a direct, or even indirect, line connecting her death to the dilemmas of water management, Prud'homme held fast to the plan of infusing his first pages with threat and dread. Launching into a series of ominous questions, he ratcheted up the terror lurking in water plant operations: "Why would someone murder a respected hydrochemist? Did it have anything to do with the quality of water at the plant?" Consider the effort that went into this next wonderful query, as overstretched in sentence structure as it is in evidence: "Were any of the more outrageous conspiracy theories—such as the claim, whispered to me in a windy parking lot, that the New Jersey mob had been angered by [the plant's] switch from chlorine to ozone treatment, a move that supposedly curtailed work done by contractors under mob control, and had put out a hit on [the drowned hydrochemist]—true?" Alas for the chance to replicate *Chinatown*'s marketing success, there is bad news ahead: "(No evidence)," Prud'homme acknowledges in an admission wrapped in parentheses to soften readers' disappointment, "(has been presented to back this theory.)"[2]

The situation in which Denver Water had landed in 1991 might seem to offer hope for the Prud'hommean literary school of resorting to murder to fight off reader ennui. A proud, powerful, and inflexible institution had been subjected to a demeaning, dispiriting, and very public defeat. An ambitious writer might reasonably hope that this situation would produce shocking plot twists that would enhance book sales. After 1990, with the plan to build Two Forks having sunk to the cold, hard bottom of the federal permitting tank, all hell would surely break loose, with hair-raising internecine betrayal and violence and bitter, ideally murderous conflict with ranchers, river rafters, and skiing industry CEOs in the stunning canyons of the Rocky Mountains. But Denver Water betrayed these expectations. Its defeat led to many meetings and negotiations, and not one documented fistfight. Not a single soul has approached me in a windy parking lot (though we have plenty of those available in this state) to whisper that the Colorado mob played a role in the doings of Denver Water. In the place of Prud'homme's unrestrained and overheated questions, the historian of Denver Water is stuck asking a prosaic, pedestrian, but actually quite important question: did the water agency truly change after the Environmental Protection Agency's veto of the Two Forks project knocked it for a loop? In 1991 Denver Water still occupied the catbird seat in the state of Colorado, securely in possession of a vast collection of water rights and of a giant, well-built network of dams, reservoirs, tunnels, ditches, pipes, and treatment plants. With these enormous assets, the agency was hardly in a precarious or vulnerable state; it could, if its leaders so chose, stick with some form of business as usual. And yet, in the words of a knowledgeable participant-observer, the veto "made the Denver Water Department go back to square one, to reexamine everything we do."[3] But was the return to square one a first step toward lasting change, or was it only a brief and temporary retreat?

In appraising the scale of change in the last twenty years of Denver Water's history, we are fortunate to turn to a thought-provoking set of reflections by water policy scholars Denise Lach, Helen Ingram, and Steve Rayner. For an article published in 2005, "Maintaining the Status Quo: How Institutional Norms and Practices Create Conservative Water Organizations," these authors studied water management in three locations: "The Columbia River system of the Pacific Northwest, the Metropolitan Water District of Southern California, and the Potomac River Basin/Chesapeake Bay in the Washington, DC metropolitan region." While we might wish that they had included the Denver metropolitan region in their study, with the size and significance of the three case studies they chose, the authors had clearly not assigned themselves light duty.[4]

Conducting many interviews to examine the capacity for change in large water organizations, Lach, Ingram, and Rayner reached a firm conclusion: these organizations were conservative, risk adverse, and committed to maintaining the status quo. The authors found abundant evidence that the continuing power of "longstanding norms and practices" produced only "timid experiments with incremental and marginal innovation." The resistance to major change, the authors explained, arose from the very nature of water as a resource. Dealing with "an unpredictable and inconveniently located resource," water-managing agencies "have evolved

Despite the veto of the Two Forks Dam, Denver and its suburbs continue to grow. In the early twenty-first century, Denver Water served 1.3 million customers in this modern mountain metro.

to attenuate or mitigate crises," creating and maintaining processes and structures that promised to "offset uncertainty, smooth out fluctuations, and make water more predictable." The determination to prevail over water's unpredictability thus locked them into "an extremely conservative approach to risk and decision making.[5]

Water managers, the authors argued, persisted in navigating by an unshakable "hierarchy of values": "reliability, quality, and low cost." Their unshakable devotion to these values put flexibility and innovation out of their reach, because "most US citizens," benefiting from a remarkably consistent delivery of water of high quality and low cost, had come to "expect water systems to be foolproof and fail-safe." Since "uniform excellence in the delivery of water is universally demanded," water agency officials had acquired "an aversion to activities that may lead to unreliable water delivery." Faced with new challenges, managers were "more likely to adopt something familiar than something new." Because utilities perform so reliably, "customers are reminded only infrequently that the resource is in any way limited." "Any adverse change in reliability, quality, or cost is likely to attract unwelcome attention from consumers, bureaucrats, and elected officials," causing "managers of well-run water organizations [to] tend to perceive any public attention as negative."[6]

In the early twenty-first century, as Lach, Ingram, and Rayner note, many commentators were "predicting and often advocating fundamental change in the way we manage water resources." But little could come from this endorsement of innovation, these scholars found, because the water organizations remained so unyielding in their embrace of the values of reliability, quality, and low cost. "Water consumers," the authors declared, "have been encouraged to believe that no behavior change on their part will be necessary." As a result, water managers operate with the conviction that "innovation cannot be tolerated if it produces even temporary declines in service to consumers."[7]

There are, after all, sound reasons to expect water utilities of our times to be tightly constrained by their historical origins. As historian Martin Melosi puts it, in the early years of building an urban water system, "a commitment to permanence . . . often locked in specific technologies and thus limited choice for future generations." Ironically, if a system was "too well built," it would "prove resistant to change." The weight of the past rests heavily on the utility managers of our times; "decisions made about [water] systems in the nineteenth century had a profound impact on cities more than one hundred years later."[8]

How does the case study of Denver Water in the last twenty years fit with the idea of an unshaken and fully maintained status quo of institutional practice? Might it be that the Two Forks veto set this agency's history apart from that of many of its national counterparts, giving it the distinctive historical feature of a severe defeat that could not be reversed, and thus cutting off the temptation to continue old, outmoded practices? This line of thought now deposits this chapter, like a foundling in a basket, on the threshold of the big question: After 1990, did Denver Water reject innovation and maintain the status quo, or did the agency achieve more significant change than Lach, Ingram, and Rayner thought possible?

⸫

For anyone interested in the future of the West's natural resources, Denver Water's actions after 1990 carried great symbolic value. As one of the nation's most focused and hard-driving agencies, charging full force at the acquisition and development of a natural resource, Denver Water's capacity to change offered a key measure of the flexibility or rigidity of the institutions created a century ago, at the height of the Progressive Era. Would these influential and powerful agencies prove to be brittle and stiff with age, more prone to fracturing than to bending? Was there unrecognized and unexplored flexibility hidden within them? Or would the results of this test of time turn out to be ambiguous, cryptic, and difficult to pin down? Might institutional response register closer to wrapping repeated assertions of receptivity to change as a veil over a determined effort to reinstall a disrupted status quo? If this last interpretation proved the most accurate, this would not necessarily indicate hypocrisy or deception. The capacity to declare a sincerely held principle and, simultaneously, to dance defiantly upon that very same principle is a well-established human talent. If the leaders of Denver Water after 1990 chose to speak eloquently of their environmental commitments and to testify earnestly to their desire to work respectfully and collaboratively with their one-time rivals, there was no particular reason to think that what they did would resemble, in every detail, what they said.

The fact that Denver Water was a relative latecomer to the popular sport of fashioning a kinder, gentler image for itself complicated the question. Learning of this late start in image cultivating is initially surprising. The controversies over the Foothills Treatment Plant and the Two Forks Dam, an outsider might have thought, would have caused the Denver Water Board to mobilize a team of public relations experts to counter the effective communication strategies of the environmental groups. On the contrary, throughout the 1980s, the utility's leaders held to the conviction that they should not spend public resources on attempts to influence the opinions of office holders or general citizens. Public Affairs was a small part of the Administration Division, dwarfed by its fellow units of Conservation, Personnel, Customer Service, Engineering Records, and Water Sales.

One of the first big, institutional changes after the Two Forks defeat was the creation of a separate division for public affairs in 1991. For readers of a skeptical turn of mind, the rise of public relations in Denver Water's operations cooks the goose. Establishing this new division gave skeptics the opportunity to conclude that the organization was now, purposefully and strategically, presenting to the world a smiley face and thereby drawing attention away from formally renounced, but energetically pursued, machinations and schemes.*

One big obstacle blocks easy access to this skeptical appraisal. Even the most creative, cunning, and clever of public relations professionals could not have designed the character traits (and certainly not the name) of the man appointed manager of Denver Water after the Two Forks veto: Hamlet "Chips" Barry III.† In his wit, self-mockery, and irreverence, Barry was unique among bureaucrats. "His favorite fish, with which he identifies closely—perhaps because of his current job," Barry said in a self-description, "is the flannel mouth sucker." A specimen "of these remarkable fish" was mounted on his office wall, "and it is the only vertebrate in the building with which his looks compare favorably." It was hard, however, to find many occasions in which Barry spoke in a

* Historians might be forgiven for thinking with nostalgia of the bluntness of Glenn Saunders, who put so little strategy into artfully designing statements aimed at disarming doubters or coaxing opponents to let down their guard.

† The effort to think of public officials who, like Barry, had senses of humor as breathtaking and unrestricted as the region's scenery has brought to mind only two counterparts: Arizona congressman Morris Udall and Wyoming senator Alan Simpson. The author would welcome other nominations.

flannel-mouthed manner. Early in his time in office, on a visit to Grand Junction, a bastion of strong Western Slope feeling against Denver Water, Barry declared that the Denver Water Department had finally left behind "its earlier adolescent personality. We're now a more mature organization." Barry was equally outspoken in acknowledging and accepting the massive change in public attitudes: "Beliefs that belonged to the environmental fringe in the 1960s have become mainstream values today."[9]

Did the Denver Water Board know what new frontiers in temperament, and even in quirkiness, they were about to explore? "It was pretty clear to me that the timeless pattern of how Denver Water did business was really not the way to continue in the future," Barry reminisced in 2009. "Clearly, if the board wanted more of the same, they would have hired more of the same. They hired me instead." Barry recognized that Denver Water had to "quit fighting with everybody" and "make some friends."[10]

To the enormous task of "changing the public face of the water department," Barry brought a very distinctive personality. In any private gathering or public event, he was a chinook of fresh air (as opposed to a mere breath). After a nearly fatal car accident, Barry lost some acuity of his hearing. He seemed to find this more a source of hilarity than misery. In a crowded restaurant, a friend told him of her gardening troubles: "We are having some troubles with the groundhogs," she said. "I think they are coming around because they like my hibiscus." "If you don't want them around," Barry replied, "why on earth are you feeding them hot biscuits?" Barry often reminisced about his many occasions of puzzlement when he entered the world of water management, struggling "to absorb the language and culture of a water utility." It had taken him at least six months to come to grips with the "use of ESP in water availability predictions." While "he thought ESP meant 'Extra Sensory Perception'" and approved of its use in attempts to glimpse the future, he eventually came to realize that ESP "is a computer acronym meaning 'Extended Streamflow Prediction.'" It is difficult to imagine a public relations professional who would have been agile and energetic enough to "manage" Chips Barry, even if that professional tried to get in shape with an intensive reading program in Mark Twain and Will Rogers.[11]

Chips Barry was both funny and serious, both fully aware of and yet undaunted by the challenges of institutional change. He

Hamlet "Chips" Barry III around the time he became manager of Denver Water.

was not an engineer, and though he was a lawyer, he had not specialized in water law and had not risen through the ranks of Denver Water. He grew up in Denver, went to Yale, and then got his law degree at Columbia. He and his wife became VISTA (Volunteers in Service to America) volunteers in rural Alaska, where Chips was "a bush attorney for Alaska Legal Services." The Barrys returned to Denver, but after a year of clerking for a judge, he headed for the Marshall Islands, responding to an invitation to work for Micronesian Legal Services, where some of his cases arose from "the consequences of US atomic weapons testing on the Marshallese people and their atolls." The route that Barry traveled to Denver Water could not be described as linear.[12]

By the mid-1970s, Barry had returned to Colorado, entered government service, and become a figure in the world of environmental policy. Starting as a consultant for the Colorado Department of Natural Resources (DNR), he became head of the Division of Mined Land Reclamation. He then became deputy director of the department in 1980, acting director in 1983, and (after an interlude at a law firm) director in 1987, appointed by Governor Roy Romer. The domain of the DNR is wide-ranging: it encompasses "the state's water policies and water projects, mining, oil and gas regulation," as well as "state lands and the state divisions of parks and outdoor recreation and wildlife." At DNR, a *Denver Post* reporter noted in 1990, "Barry's approach has been to keep his political cards close to [his] chest, bring together feuding parties and negotiate." As director of DNR, Barry was well-positioned to observe the struggle over the Two Forks Dam, with ample opportunities to study the Denver Water Board's customs and conduct. While he was still at the DNR, Barry accompanied Governor Romer to meet with the principal figures at Denver Water. The governor asked the group a direct question: "I want to know what you're going to do if Two Forks doesn't get built. What's your plan B?" Each person around the table offered more or less the same answer: "We don't have a plan B." Not long after, Barry held the job of Denver Water's manager, with a rare opportunity to realize how accurate the answer to the governor's question had been.[13]

As much as the choice of the new manager, the appointment of Denver Water's new lead counsel in 1991 symbolized the scale of change in the agency. Taking the position once held by the legendary Glenn Saunders was Patti Wells (unlike Saunders, she would not be affiliated with an outside law firm, and Denver Water was the only client she would represent). As much as Barry's, the career track leading Wells to Denver Water had elements of surprise: she was a specialist in environmental law, with her JD from Harvard, who had spent a good share of her career as an attorney at the Environmental Defense Fund. In another big change in the cast of characters, Denver mayors also took to appointing to the Denver Water Board the sort of people who, in an earlier era, might have been its firm opponents. Serving from 1995 to 2007, Denise Maes was a specialist in environmental law. Joe Shoemaker, in office from 1995 to 2001, was the leading figure in the restoration of the South Platte River in its course through the city. George Beardsley, a developer and forceful thinker who took up many environmental causes, serving on the boards of the Great Outdoors Colorado Trust Fund and the Nature Conservancy, was on the Denver Water Board from 2004 to 2009. Susan Daggett worked for years as the lead attorney for the Denver office of the environmental law firm Earthjustice before she joined the board in 2007 (she stepped down in 2009 when her husband, Michael Bennet, was appointed to the US Senate).

If we shift back to the frame of reference held by the principal figures of Denver Water only a few years before, it might well have seemed that the barbarians were no longer at the gate; the gate had swung open, and the barbarians now occupied the central offices in the fortress. The change in personnel extended well beyond the main offices. Before 1991, the custom of hiring from within had produced conditions that fostered a resistance to change. "If you asked the question, 'Why do we do it that way?' the typical answer," Barry remembered from his early years at Denver Water, "was 'We've always done it that way,'" an answer he found "incredibly unpersuasive." One policy that contributed to the insular culture at Denver Water was the practice of limiting applications for many job openings to current employees. In an innovation that Barry saw as essential for changing the institution's culture, he put in place an open and competitive hiring process and began regularly advertising jobs outside the agency.[14]

In their prior careers, in their personalities and styles of communication and conduct, and in their stance toward the environmental laws and regulations of the federal government, the new leaders of Denver Water differed significantly from the old leaders.

Or did they?

For all the factors of change, Denver Water's new leaders did not waver in their commitment to secure an adequate supply of water for the residents of the City and County of Denver, which was, after all, the charter mandate they accepted when they took these jobs. Denver Water's new leaders were as committed as their predecessors to the practice of taking foresighted action in the present to avoid shortages in the future. And, perhaps most to the point, in the state of Colorado, Denver Water still held a tremendous advantage in senior water rights and in the ownership of a massive infrastructure for capturing and distributing water. Thus, declarations of a new era of congenial collaboration had the power to arouse the doubts intrinsic to situations when a five-hundred-pound canary takes to chirping pleasantly about respecting the needs and sensitivities of tiny finches and sparrows unfortunate enough to be located in the neighborhood of this canary's massive bird feeder.

How did the status quo fare in the 1990s and early 2000s? To what degree does the answer to this question ratify—or challenge—the conclusions offered by Lach, Ingram, and Rayner on the maintenance of the status quo?

♦♦♦

The practice known as demand management offers a compelling challenge to the judgment that Denver Water could not budge from the status quo. Rather than assume that people will keep using a resource or commodity (energy, food, wood, water, etc.) at a steady or accelerating rate that cannot be changed and must simply be served and accommodated, demand management liberates organizations and communities from passivity and fatalism. People may have convinced themselves they cannot possibly get by with less in the way of a particular resource, and the officials of the agencies that serve them may think that there is nothing to do but meet their persnickety clients' demands. The techniques of demand management repeatedly demonstrate that people can get by with less. If you place Denver Water's performance in demand management next to the assertions (by Lach, Ingram, and Rayner) that "customers are reminded only infrequently that the resource is in any way limited" and that "water consumers have been encouraged to believe that no behavior change on their part will be necessary," those assertions make a strikingly poor fit to Denver's story.[15]

In fact, the history of Denver Water's support for reducing water use scrambles any effort to present a clear-cut narrative that would be marked by a sharp distinction before and after 1990. As early as the 1920s, periodic initiatives had encouraged citizens to avoid wasting water. In a provision of the Foothills Agreement in 1979, Denver Water "had to commit to implementing a water conservation program," with specific "targeted levels" to meet. As early as 1981, Denver Water had made its mark in conservation by coining the term *Xeriscaping*, combining *landscape* with *xeros*, the Greek word for "dry," and popularizing the practice of adorning yards with plants that required comparatively little water. "Xeriscape," as Denver Water's website defines it, "uses low-water-use plants to create a landscape that is sustainable in Colorado's semi-arid climate."* In 1996, Denver Water and the American Water Works Association collaborated to publish a colorful and persuasive book that clearly showed how a properly designed Xeriscaped garden could be as beautiful as one upon which water had been profligately dumped. "Why, as gardeners, should we torture ourselves, growing plants from radically different climates?" Rob Proctor wrote in the book's introduction. "Why squander precious water on exotics that stand a slim chance of survival or that never fulfill their promise?" The book very directly challenged the once-dominant aesthetic of greenness: "Green lawns and water-guzzling ornamental plants seem out of place in a semiarid environment."[16]

As a prosaic but consequential foundation for conservation, the installation of water meters had begun years before the Two Forks veto. Residents cannot, after all, make much progress in the project of reducing the scale of their water use if they do not know how much water they are using. If they are paying a flat rate and no meter registers their actual usage, they are off to a bad start in improving their habits. In 1957, Denver Water began requiring the installation of water meters in new homes within the city. Pre-1957 homes stayed with flat-rate service until 1986, when a project

* This word has had a tough history, with frequent failures of enunciation changing it from "Xeriscape" to "Xeroscape," with the next stage of slippage leading to "zeroscape." Even the well-intentioned can thereby erase the richness, complexity, and beauty of a Xeriscape by punning it into a place characterized by nothing.

to convert those homes to meters was launched. By 1992, Denver Water finally had meters for all homes within its service area, making it possible to charge users a fee proportionate to their actual consumption, a crucial condition for encouraging conservation.[17]

An even bigger step, begun as the campaign for Two Forks approached its end, was the creation of increasing block or tiered rates, by which the price for water climbs proportionately to the amount of water used. Proposed in 1989 and put into practice in 1990, block rates marked Denver Water's emerging commitment to use residents' water bills to modify their habits of water use. "Our customers," Chips Barry wrote in 2005, "have expressed a desire for water rates that are higher for those who use water inefficiently while more directly rewarding efficient use."[18] If bartenders applied this method to the cause of reducing people's consumption of beer, they would charge one dollar for the first beer, two dollars for the second beer, four dollars for the third beer, eight dollars for the fourth beer, until the price reached a point when even those drinkers whose judgment had become clouded would come to see the fiscal argument for reducing their consumption. With this mechanism, Denver Water was actively and purposefully using pricing to change customer behavior, to promote conservation, and to reduce waste. In their article, "Maintaining the Status Quo," Lach, Ingram, and Rayner had placed a determination to provide water at a low cost among the three values that water utilities would not rethink or reject. The higher rates in the increasing block or tiered system showed the agency's willingness to reconsider the assumption that its rates must, always and invariably, stay as low as possible.

In another direct approach to changing behavior, Denver Water under Chips Barry's direction pushed the cause of conservation with advertisements that, in their madcap style, reflected the fact that a bureaucrat with an unusually vigorous sense of humor was calling the shots. Hiring a small local advertising company, Sukle Advertising, Denver Water responded to the severe drought of 2002 by jettisoning the usual earnest and dull genre of exhortation to more responsible resource use. Eccentric slogans—"Instead of a Dishwasher, Get a Dog" or "Only Wash the Stinky Parts"—appeared on sandwich boards and on billboards. At certain major sporting events, a person costumed as a toilet would dash across the field with security personnel in hot pursuit (usually ending the chase with a hard tackle) while the crowd was admonished "Don't let your toilet run." In 2006, Sukle returned with a campaign centered on the suggestion "Use Only What You Need" displayed on billboards and benches from which much of the wood and framing had been removed, leaving only the area needed to display the slogan. One of the most unconventional of Denver Water's pitches played on masculinity. The alarming observation that "Real Men Dry Shave" was distributed on coasters, while extraordinary radio ads—like the one quoted at the start of this chapter—encouraged men who were experiencing anxieties about their masculinity to resolve these tensions by adjusting their sprinklers. Original and lively advertisements, supplemented with pricing that discouraged excessive use and with various efforts to promote more efficient household appliances, had a remarkable impact on levels of water consumption. Before 2002, the average per-person-per-day water use registered at 211 gallons. In 2010, customers were using 171 gallons per person per day, a 19 percent drop from predrought levels.[19*]

Proven in practice, demand management could then be incorporated into Denver Water's projections of future demand and supply. As Barry described the procedure (admitting to oversimplifying it), in previous times "you knew what your future demand would be because you projected your future population; and if it went up 30 percent your demand went up 30 percent, as a simple extrapolation." Rather than estimating future need for water by anticipating a certain growth rate and assuming an unchanged per capita rate of water usage, Denver Water could now purposefully reduce the demand and also add conservation gains to the projected supply. Water saved through conservation could now be considered as nearly equivalent to a new source of supply and calculated into projections of future resources. Every drop of water redeemed from waste through conservation is, in essence, a new drop of water, an enrichment and enhancement of what the city's residents have on hand to use for their actual needs.

As the agency's innovative Integrated Resource Plan (IRP) put it in 1997, conservation would "become an integral part of Denver's

* These use volumes reflect a crude calculation of Denver Water's total water use divided by the total population in the agency's service area. As such, it includes water used by industry, water used by tourists, and water used at venues such as sports stadiums, the performing arts complex, and Denver International Airport. As a result this per-person-per-day measure records a higher figure than the actual consumption by the average individual Denver Water customer.

Denver Water's innovative advertising campaigns in recent years have reminded customers to use only what they need and to not let toilets run.

new supply equation, not just a contingency in times of drought." By encouraging Xeriscaping, metering use, setting tiered rates, and appealing to customers' senses of humor (and even their appreciation of absurdity), Denver Water worked to make good on the IRP's claim that "now conservation is considered a significant means to develop new 'sources' of water supply." Very much in contrast to the conclusion reached by the team of Lach, Ingram, and Rayner, portraying water management agencies as incapable of challenging consumer expectations, Denver Water's leaders had explicitly and intentionally moved into the domain of changing attitudes and conduct: "Getting people to conserve water in their homes and businesses," the agency declared publicly, "involves attempting to modify their behavior and habits."[20]

And did successful water conservation add up to the one arena of simplicity in the tangled world of water management? Was this the one initiative that could earn Denver Water a round of unanimous applause from every possible stakeholder? On the contrary, successful conservation carried a lasting paradox in customer relations. When Denver Water persuaded residents to conserve and to avoid waste, success led to a drop in the utility's income as customers bought less water. And yet the fixed costs of maintaining and operating the utility's infrastructure did not drop. Thus, Denver Water had to respond to this decrease in revenue by raising its rates. Water users had risen to the challenge and embraced the practices of water conservation, and then they were "rewarded" by having their rates raised. Customers thought their good efforts at conservation should reduce the amount they were charged, but Denver Water's officials still had to make up for the drop in revenue that their efforts at persuasion had produced. Even an agency that had been cultivating its skills in communicating with the public was hard put to get this message across, despite earnest and accurate statements that "fixed system costs remain relatively the same over the short term, however much or little water customers use."[21]

Along with demand management, the designation of a combined service area (CSA) represented another big element of institutional change. As the organization's IRP put it in 1997, "With the veto of Two Forks, it became clear that the Board could no longer take upon itself responsibility for the water supply future of a rapidly expanding Denver Metro Area." As Barry said, "We didn't have any boundaries on our service area." Instead, Denver Water had "held the door open and said, 'If you want water from Denver, come talk to us and maybe we'll give you a contract.'" The creation of the CSA, "consisting of the City and County of Denver plus 75 contractual distributors in its suburbs," ended that much more flexible era. As of 1994, the water suppliers within the CSA, as holders of "new distributor contracts," had the promise of service to build out, while those outside the CSA, even if they held contracts with Denver Water for a fixed amount of water, did not receive any such commitment regarding their future needs. "Providing for the service area" would be "the foremost obligation of Denver Water. Only secondarily would it look outside its service area boundary for potential efforts which might prove mutually beneficial to both Denver's service area and metro regions beyond."[22]

The creation of the CSA had not, however, fully resolved the question of the relationship between Denver Water and the communities that fell outside the service area. "Clearly Denver Water does not need to accept responsibility for meeting the water needs for exploding Metro Denver growth from the El Paso County line on the south to the Northern Colorado Water Conservancy district boundary on the north," the IRP stated in 1997. But that did not close the question: "Yet, the Board recognizes that it is not an island—that it is economically and socially integrated to a larger metro area, much of which exists outside its service area." That larger terrain was "linked to the future fortunes of the City of Denver," making it difficult, if not impossible, for Denver Water to chart its course in a detached and independent manner.[23]

To the south of Denver, this conundrum presented itself sharply. The population of Douglas County had remained small through much of the twentieth century. In the 1980s and 1990s, its open spaces attracted the attention of developers, and the county emerged as one of the fastest growing counties in the nation. In 1970, Douglas County had a population of a little more than 8,000; by 2000, the county's population reached 175,766; and by 2010 the population had grown to 285,465.[24] Douglas County had, however, made a very late start in acquiring water rights to the state's rivers, Front Range or Western Slope. Instead, the expanding communities of the county relied heavily on nonrenewable groundwater, and the pumping of aquifers could only be a short-term solution.*

* In this approach, Douglas County of course replicated the historical nationwide pattern,

Until very recent times, an observer might have thought that Douglas County and the City and County of Denver had found their role models in Aesop's fable of the grasshopper and the ant. Denver Water had labored away, exercising foresight and preparing for seasons of scarcity. Meanwhile, in the manner of the grasshopper, the governments, developers, and new residents of Douglas County were "hopping about, chirping and singing to their hearts' content" and finding no inspiration in all the institutional "toiling and moiling" just to the north. But finally Douglas County got the point: "It is best," as Aesop's moral to the story puts it, "to prepare for the days of necessity."[25] Denver Water had thought ahead, acquired water rights ahead of the game, and labored relentlessly to build the structures to store, divert, and manage its resources. Douglas County, comparatively late in the game, had plunged into a festival of growth and development and now eyed the results of its neighbor's earlier "toiling and moiling" with yearning.

Should the security earned by foresight be reallocated to rescue the not-so-foresighted? Denver Water did not have "any legal obligation to serve anyone" outside the CSA, Barry said in 2008. But that was not the end of his line of reflection: "That's one of my big problems—we don't have a legal obligation, but we might have a moral and a political obligation."[26] There would be no advantage to Denver in a grim, well-publicized scenario in which a neighboring county, providing homes to many of the people who worked, shopped, and enjoyed the attractions in the metropolis, plunged toward a difficult future of water scarcity. When the finite groundwater supply became more and more expensive to secure and homeowners had to confront the precarious circumstances of their water supply, the blow would land on the reputation of the whole metropolitan region.

Douglas County communities and residents, meanwhile, undertook some striking changes in their assumptions and practices. An unusual set of gatherings, featuring remarkably frank discussions of the county's water supply, began to occur. Programs in conservation and Xeriscaping proliferated, with high school students trained and deployed to install water-efficient sprinkler nozzles for yards. Highlands Ranch—once featured in *National Geographic* as the epitome of suburban sprawl—began creating water budgets

relying on wells in the first phases of growth.

for every customer in 2003, with water bills informing customers how much water they should use each month. Douglas County's residents beat the residents of Boulder, Colorado, one of the environmentalist centers of the West, in reducing its water demands. In surveys conducted by the county government, 81 percent of citizens said that they believed "water to be the biggest challenge faced by the region." Discussions exploring the possibility of reuse, treating water that had already traveled through the distribution system of Denver Water and the large suburb of Aurora, began to gain momentum, signifying a shift away from confrontation toward negotiation in the relationship between Denver Water and the suburbs to the south. (In a rare case of acronym success, the water-sharing arrangement under discussion is called the Water Infrastructure and Supply Efficiency project, or WISE).[27]

In his dealings with Denver's suburbs and with the Western Slope, Barry made a number of overtures toward direct discussion and negotiation, seeking out arrangements that might benefit both Denver and its one-time rivals. One example appeared in the arrangements with the Western Slope over the proposed Wolford Mountain Reservoir, to be built by the Colorado River Water Conservation District. When he took office, Barry did not find the existing arrangements for this reservoir to be very attractive: Denver Water was going to pay part of the reservoir's construction cost in return for a lease that would secure roughly half of the water for twenty-five years. That twenty-five-year time frame would have supplied water to the Denver Water system during the construction of Two Forks Reservoir, but after Two Forks was vetoed, the temporary lease lost much of its attraction for Denver Water. Meanwhile, the Colorado River Water Conservation District struggled to raise its share of the money to complete the reservoir. Barry proposed a different arrangement: in return for Denver Water contributing most of the money needed for the completion of the reservoir, it would have a lasting claim on 40 percent of the water (to be used as compensation water, not for diversion to the Front Range), rather than a transitory twenty-five-year lease. Wolford Mountain was thus redefined as a joint-use structure, with benefits shared by both the Western Slope and the Front Range.[28]

In the mounting recognition that many of the "additional sources of secure supply" were already claimed and developed, scholars Lach, Ingram, and Rayner saw the conditions for conflict and contestation. Current circumstances, in which there are comparatively few "further additional sources of secure [water] supply," the authors wrote, would result "in agencies colliding and struggling over control over increasingly scarce water supplies"; "collaboration among water agencies becomes difficult because sources of additional supplies that promise benefits to all participants become increasingly difficult to identify." In its negotiations with suburbs and with the Western Slope, Denver Water presents a case study that challenges this prediction of universally strained collaboration.[29]

In a comparable demonstration of adaptability, Denver Water led the region (and arguably the nation!) in a forthright recognition of and reckoning with the reality of global warming, staying out of the swamp of partisan stewing over whether global warming was a hoax perpetrated by scientists cleverly stirring up alarm to generate funding for their research. "I don't think there's any question that global climate change is real," Barry said. "There is plenty of evidence that the world is getting warmer."[30] Accepting the reality of global warming meant confronting a world in which the usual methods of prediction had lost their relevance. The old habit of planning for the future by examining the historical record of precipitation, snowpack, and stream flow had been rendered a matter of quaint and antiquarian curiosity. With climate change, there were few lessons to be gleaned from attending to the patterns of the past. Moreover, there was no certain way to translate the broad, global transformations of climate into site-specific, exact predictions of precipitation. Some areas would be wetter, and some would be drier, and good luck to the brave souls whose jobs required them to squint into the future to predict which would be which. Barry spoke forcefully about the uncertainty water agencies faced: "The problem is that nobody knows what a warmer climate does to the pattern and volume of precipitation. I fear that people think we're moving from climate A to climate B. The fact is that we may be moving from climate A to a continuously shifting climate. It's not like we're going to reach a point and then we'll know what the new climate will be. It's a shifting target. About the time that we figure out what the shift is, it shifts again."[31]

In placing concern over the impacts of global warming at the center of its deliberations on the future, Denver Water had embraced a central tenet of environmentalist thinking. And yet the consequence of that agreement was ironic: *not* putting up a fight against environmental concern about global warming led Denver Water to propose one more big storage project that would trigger opposition from some environmental advocates. Since global warming meant greater unpredictability in the future scarcity or abundance of water in Colorado, foresight seemed to require Denver Water to diversify the sources of supply and to build greater resilience and flexibility into its system. As Barry often said, an organization like his was now engaged in managing risk as much as managing water.

💧💧💧

Whatever else had changed at Denver Water, the determination to look ahead, to anticipate trouble, and to take actions in the present that would keep an adequate supply of water traveling in the pipes to the customers and clients had not diminished. This is the territory where the conclusions of our oft-quoted scholars Lach, Ingram, and Rayner, make the closest match to Denver Water: water managers did indeed put reliability at the top of their "hierarchy of values." As the IRP of 1997 put it, "Since in the West a water shortage is likely to affect many water suppliers simultaneously, there is no quick bailout if planning has fallen short." Even as the plan had awarded an enhanced status to water conservation in the near-term, it did not foreclose the possibility of the construction of storage structures in the long-term. "The Board cannot be certain that conservation, reuse, and system refinements will suffice in meeting all of Denver Water's future demand," the plan stated. "If they do not, the Board must reserve its water rights so that future storage efforts can be undertaken."[32]

In 2002, a severe drought rattled the confidence and undermined the assumptions of everyone involved in water management in the state of Colorado and in the Colorado River Basin. In the late spring, Denver Water's reservoirs were 73 percent full. In early July, the percentage had dropped to 66 percent. By mid-July, Cheesman Reservoir was only 32 percent full. In September 2002, Denver Water's reservoirs were 51 percent full, and in the spring of next year, that percentage had dropped to 44 percent. Even though spring runoff quickly relieved this grim situation in 2003, the drought nonetheless lodged as a constant presence in the minds of everyone who worked at Denver Water, making it untenable to think of intense drought as a remote possibility, statistically improbable and exaggerated by alarmists. Along with disturbing reductions in the reservoirs' holdings, dry conditions also led to a frightening season of wildland fire. The forests in Denver Water's watersheds along the South Platte were left vulnerable to erosion, and sediment and ash accumulated in Cheesman and Strontia Springs Reservoirs.*

Relying heavily on Denver Water's southern system (the South Platte, along with the diversions into the South Platte from the Western Slope through Roberts Tunnel) now seemed like a precarious and worrisome practice. Haunted by their experiences in 2002 and anticipating an uncertain future with climate change, Denver Water's leaders declared an intention to strengthen its northern system, by diverting more water from the Western Slope through the Moffat Tunnel and increasing the storage capacity of Gross Reservoir on the Front Range to hold the water so obtained.

In 2003, Denver Water filed for a permit to expand Gross Reservoir on the Front Range, raising the dam by 125 feet. The project would nearly triple the storage capacity, from 41,811 acre-feet to 114,000 acre-feet. Obtaining more water to hold in the reservoir would entail increasing Denver Water's diversion from the Fraser and Williams Fork Rivers on the Western Slope through the Moffat Tunnel. This proposal had an interesting history: environmental groups opposing the Two Forks Dam had offered the Gross Reservoir expansion as an acceptable alternative, and a 2003 report from an environmental consortium called the Sustainable Water Caucus mentioned the expansion as an idea worth consideration. This report set out the context in which a project in increased Front Range storage must be pursued: "Ultimately, Denver should proceed in a manner that addresses Fraser River basin water supply and instream flow problems as well as Front Range instream flow problems in cooperation with local water users in each of these basins."[33]

In presenting the plan to the public, Denver Water's leaders declared that they would secure half of the water they would need by 2030 through conservation. But the other half would come from expanded storage. Before any construction could begin, in the procedure familiar from Foothills Treatment Plant and Two

* The problem was most acute in the much smaller Strontia Springs Reservoir.

The Hayman Fire, which burned in the South Platte River Basin in June 2002—right up to the shores of Cheesman Reservoir, as this photo shows—was the largest wildland fire in Colorado's history. In the aftermath of the fire, Denver Water has battled erosion and the accumulation of sediment in reservoirs below the burn area.

Completed in 1954, Gross Dam stores nearly 42,000 acre-feet of water in Denver Water's Moffat Collection System. In 2003, Denver Water filed for a permit to raise the dam 125 feet and nearly triple the reservoir's storage capacity to 114,000 acre-feet.

Forks, the US Army Corps of Engineers would have to prepare an environmental impact statement and then decide whether to issue a 404 Dredge and Fill Permit. Issued in October 2009, the corps' draft environmental impact statement (DEIS) set off a round of protest, voiced in a variety of tones and styles.

Affiliating with several groups (the Environmental Group of Coal Creek Canyon, Coal Creek Canyon Improvement Association, Friends of the Foothills, and concerned citizens of the Gross Reservoir community), landowners in the vicinity of Gross Reservoir issued a spirited statement of opposition condemning the corps' DEIS on a battery of counts: it did "not adequately portray the long-term and irreversible damage to the local and Western Slope environments that the project would introduce"; it failed "to examine the rationale for the proposed project carefully"; it did "not consider the environmentally non-invasive alternatives for effectively addressing the future water needs of the Denver region." The neighbors' group condemned the "further depletion of Western Slope rivers," expressed dismay over the "long-term and irreversible damage to the environment and ecosystems" of their own neighborhood, and then notched up the heat on their rhetoric: "Are we not at the dead-end of a disrespectful and morally tone-deaf attitude of aggression, rapaciousness, and selfishness toward living with an interconnected and interdependent matrix of life?" This stem-winder of a DEIS public response continued, "Why should we prolong this attitude just one more time via the Gross Reservoir expansion project when much damage has already been imposed on our physical environment and our moral fiber?"[34]

For much of the last half century, a significant sector of Americans condemning the environmental impacts of resource development have been forgiving of, if not oblivious to, their own resource use. Residing on the high ground of the foothills and mountains, the people living in proximity to Gross Reservoir were not on comparably high ground when it came to sustainable practices touching lightly on the earth's natural systems. The impact of exurban residential sprawl on "our physical environment and moral fiber" bore an uncomfortable resemblance to, or even kinship with, the operations of Denver Water. Even as they excoriated the failure of Denver's citizens to give up their wicked ways and embrace environmental principles, the reservoir neighbors were themselves residents of an area of exurban settlement where wells drawing on groundwater-supplied homes located in sensitive wildlife habitat. These homes were located in a forest subject to periodic wildfires under natural conditions; in similar locales in the Rockies, the deployment of young men and women as firefighters, as well as big expenditures of federal, state, and county resources, came to the rescue (not always successfully) of these vulnerable homes. Residents came and went from their Edenic setting, thanks to the intensive use of fossil fuel–powered vehicles traveling on state-subsidized roads that, in themselves, constituted something of an intrusion into the "interconnected and interdependent matrix of life" for which Denver Water was exhorted to feel more respect.

Meanwhile, in a more temperate tone, acknowledging their appreciation for ongoing discussions with Denver Water and complimenting the organization for its innovations in conservation, the organizations Trout Unlimited, Western Resource Advocates, and the Colorado Environmental Coalition* focused their concern on the impact of increasing the diversion from the Western Slope to secure the water that would be held in the enlarged Gross Reservoir. Their comments on the corps' DEIS concentrated on the troubles of the Western Slope's Fraser River, already significantly dewatered by preexisting diversions. The phrase *cumulative impacts,* seemingly humdrum but full of consequence, was a key to their argument. Denver Water's diversions had already reduced the Fraser's flow by half,[35] and Trout Unlimited and Western Resource Advocates raised the prospect that a further reduction in flow would cross a threshold or tipping point, with an escalating decline in the river's ability to support fish and the aquatic organisms on which fish depended. And the corps' DEIS, they said, had not reckoned with the impact of a contemporary project, the effort of the Northern Colorado Water Conservancy District to expand its own Front Range storage for diversions from its Windy Gap transbasin system, which would create yet another cumulative impact on the watershed of the Colorado River. Placed in the full context of both existing and planned diversions, the Moffat System Expansion had

* The comments were submitted on behalf of the Colorado Environmental Coalition, Western Resource Advocates, The Nature Conservancy, Center for Native Ecosystems (now Rocky Mountain Wild), San Juan Citizens Alliance, Clean Water Action, Save the Poudre, and Sierra Club–Rocky Mountain Chapter. Another major document is the joint rebuttal report submitted by Grand County, Summit County, Northwest Colorado Council of Governments, Middle Park Water Conservancy District, Trout Unlimited, Colorado River Water Conservation District, and Western Resource Advocates.

to take its place in a broader picture of previous degradation of the Fraser. This, the project's critics argued, made it necessary for the corps to expand its attention to the prospects for the reduction of the peak flows needed to flush out sediment, for summer water temperatures heightened by the constrained flow of water in the river, and for negative consequences in water quality. An equally strong line of criticism questioned the corps' screening criteria for deciding which alternatives deserved consideration and which should be summarily dismissed.[36]

The corps, these commentators asserted, had failed to put Denver Water's four justifying needs (reliability, vulnerability, flexibility, and firm yield) through a sufficiently critical appraisal. Conceding too much to Denver Water's characterization of those needs, the corps, in the judgment of Western Resource Advocates, failed "to evaluate the potential for future conservation to reduce demands."[37] Denver Water's projection of future need, opponents said, was outdated since it did not take into account its own remarkable performance in furthering conservation and reducing demand since the 2002 drought.

This controversy placed Denver Water and its opponents on opposite sides of a canyon of differing interpretation of the meaning of the 2002 drought. To the utility's leaders, the historic episode of the drought delivered an urgent mandate to seek a greater supply of water in order to be prepared for a future time of shortage. To the agency's critics, the same story carried a reassuring and encouraging message to trust to the capacity of Denver Water and its customers to change their habits and restrict their usage in times of scarcity; in other words, there were effective ways to cope with drought besides having more water in storage. Just as striking was the difference in interpretation of the implications of taking climate change seriously. The opponents criticized the DEIS for a failure to reckon with the impacts of climate change, especially in the increasing vulnerability of the Fraser River to catastrophic change from further diversion. Meanwhile, Denver Water stuck with its own characterization of climate change as a variable of enormous importance, requiring foresighted action to minimize risk in an unpredictable world.

Echoing with the language and emotion of many earlier environmental conflicts, this controversy did present one novel element: the opportunity for the project's opponents to argue that Denver

Water's extraordinary success in conservation made it unnecessary to expand any storage facilities or to divert any more water from the Western Slope. The opponents of the project regularly complimented Denver Water for its impressive achievements in water conservation, arguing that the agency's own fine performance in reducing waste proved that no new storage was necessary. The Gross Reservoir neighbors expertly delivered this combination of flattery with scolding: the corps' DEIS, they wrote, "fails to reflect the conservation savings that Denver Water is already achieving through its uniquely thoughtful management."[38]

"We are not endeavoring to be naive and strident neighborhood environmental activists," the neighbors of Gross Reservoir wrote in their statement of opposition to the project, explaining that they were, instead, viewing the project "from an overall societal, political, human and ethical standard."[39] And yet their proximity to the site was, understandably, the key to their concern. Who, after all, could expect that a group of residents would choose a site because of its natural quiet and beauty, invest money in buying land and in building pleasant homes, and *then* welcome trucks, heavy equipment, workers, roads, dust, and disruption to the neighborhood? It would also be peculiar and puzzling if the Boulder County Commissioners, or the members of the state legislature representing this area, supported a big and disruptive construction project in their own territory when that project would never deliver water to the residents of the county in which it was located. And the idea that Trout Unlimited, a remarkable organization pushing for healthy streams and optimal conditions for fishing, would support the further dewatering of an already stressed river also lands far beyond probability.

The people who oppose the expansion of the Moffat Collection System and of Gross Reservoir doubt that the scale of change in the fundamental institutional character of Denver Water has added up to much. To the cynical, the last two decades of Denver Water's activities had been a demonstration of how to put on a show of changing an agency while only redesigning its veneer. The Board Resource Statement of 1996, orchestrating a fundamental shift in the agency's agenda, declared that "the Board will pursue resource development in an environmentally responsible manner." The board "will always proceed in an environmentally sensible matter," the IRP proclaimed in 1997. Furthermore, "For Denver

Water, that guideline has become axiomatic. It believes that in all of its actions, good environmental practices are good public and water policy." More recently, the strategic plan adopted on March 9, 2011, declares that the organization will "be environmentally responsible in delivering on our mission."[40] To opponents, juxtaposing declarations like these to the plan to expand Gross Reservoir only shows Denver Water to be up to its old tricks, this time with a public relations division in full swing.

At this point, the armchair quarterback begins to shift uncomfortably in her armchair. A memorable story told by a professor back when I was in graduate school comes, uncomfortably, to mind. It seems a marriage counselor, working with a troubled couple, was going to be observed by a student in training. The student sat off to the side, and the marriage counselor brought in the wife. He listened intensely and sympathetically to her story, a tale driven by the certainty that her husband was the originator of the problems in the marriage, and when she finished, the marriage counselor nodded sagely and said, "I think you're absolutely right." The wife left, and the husband came in and launched into an opposite version of the story, stressing his innocence and his wife's status as the source of trouble. At the conclusion of the husband's narrative, the marriage counselor again nodded sagely and said, "I think you're absolutely right." With the husband out of the room, the marriage counselor turned to the student and said, "What do you think of my technique?" The student was, by this point, quite stirred up. "Your technique," she said, "is dreadful. By validating two opposite stories, you have made a bad situation even worse." The marriage counselor listened closely, nodded sagely, and said, "I think you're absolutely right."*

Striving for a level of analytic exertion and rigor a notch or two above the marriage counselor's, I offer a somewhat more precise summary of the current dispute. I can understand why the Denver Water Board wants to prepare for global warming and for episodes of drought by enhancing the northern section of its system and increasing its storage on the Front Range. In this cause, the virtues of expanding an existing dam, rather than transforming a pristine canyon into a new reservoir, are unmistakable. I can understand, moreover, why Denver Water's leaders cannot make

* I am in debt to Professor Kai Erikson at Yale University for this story.

their peace with the risks of heading into an uncertain future with water conservation savings as the major new source of supply. I can understand why homeowners in the vicinity of Gross Reservoir feel afflicted by the proposal of a major construction project. I can understand why proponents of recreational fishing and the aquatic well-being of the Fraser River fear the effects of further diversion from the stream. With the Fraser River's flow already significantly reduced from previous diversions, the dismay felt by the members of Trout Unlimited is entirely understandable. And I can see why Trout Unlimited points out the need for more expensive remediation efforts to keep a further dewatered river from reaching temperature heights in midsummer that would jeopardize aquatic wildlife.

I can also see why opponents of the current project doubt the authenticity and depth of Denver Water's declarations, over the last twenty years, of concern for the environment. While it is demonstrable that Denver Water's leaders have changed their agency's conduct to take environmental concerns into account, it is also evident that interest in the well-being of aquatic organisms and local fly-fishing economies has not displaced or eroded the commitment of these leaders to provide reliable water service to their users. They see increasing the capacity of the Moffat Collection System as essential to honoring this commitment. They do not concede the point that their determination to provide reliable water supply has undermined or overruled their environmental commitments. In fact, the agency's leaders feel that their plan to expand Gross Reservoir comes well-accompanied with commitments to provide remediation, and even enhancement, to the natural systems involved.

So what are we to think?

In the early twenty-first century, phrases like *environmentally responsible* and *good environmental practices* run rampant in the world. Hundreds of individuals, groups, and organizations distribute, circulate, and consume these phrases at a fevered pace. No one holds the power to determine the clear, universally recognized meaning of these phrases, while every group holds the power to deny their opponents the right to use them with legitimacy.

Meanwhile, every rhetorical squabble over who holds the high ground in environmentally responsible conduct presents the opportunity for the participants to contemplate the terms of their conflicts in more down-to-earth, honest, and productive ways.

Racing to occupy the two familiar poles of opinion in these disputes has become a sterile and empty performance. The advocacy of conduct in which human beings inhabit the earth so lightly that nature remains pristine and entirely undisturbed has come to seem utopian and improbable. In exactly the same way, the justifications for conduct in which human beings ruthlessly seize the earth's resources have frayed and snapped. In this historical instance, Denver Water's position—attempting to make a major shift in its practices while also carrying with it the burden of a weighty history—is an illuminating case study in a much bigger, worldwide effort to explore the many places on the spectrum between the ideal of an ethereal and insubstantial human impact on nature and the fatalistic acceptance of the hard usage of the earth and its resources. Denver Water's leaders, in other words, are searching for a zone of compatibility between honoring the mandate from their charter and honoring their considerably more recent commitment to good environmental practices. Observers of this effort are certainly entitled to perform some hand-wringing and lamentation over the unevenness of the outcome. But observers also have good reason to acknowledge—and to encourage—the positive aspects of the undertaking.

In their statements of opposition, the neighbors affected by Gross Reservoir's expansion accent that they are pressing the case for the recognition of limits. Over the better part of a century, they observe, Denver Water's relentless acquisition of water has permitted and encouraged the expansion of population and development in the metropolitan region. This process of growth, they assert, cannot go on forever. At some point in the twenty-first century, the population in Denver must stabilize, and Denver Water's leaders must realize that the time has come to cease building more structures and diverting more water from the Western Slope. Denying a permit to expand Gross Reservoir thus figures, in this vision, as the necessary and unavoidable way to convey this message to an agency that, for much of the last century, conducted itself as if restraint and limitation had no bearing on its operations.

Here is the curious twist: to a striking degree, this message on the existence of constraints that require a fundamental shift in the practices of water management in Colorado is one that Denver Water itself has been proclaiming.

In 2011, the agency's willingness to lay out and consent to limits in its future became even more striking. "After five years of mediated negotiations," Denver Water and a group of Western Slope organizations announced the Colorado River Cooperative Agreement. The representation from Western Colorado included county and city governments, irrigation and water districts, companies (including ski resorts), as well as the powerful Colorado River Water Conservation District. A briefing summary of the agreement describes it as beginning "a long-term partnership between Denver Water and the West Slope," setting up "a framework for numerous actions by the parties to benefit water supply and the environment on both sides of the Continental Divide." In the recognition of limits, one provision offered the key signifier of change. In a repeated phrase, the summary declares that, when it comes to building waterworks infrastructure on the Western Slope, "Denver Water will not undertake any future water development activities without the prior approval" of Grand County, Eagle County, Summit County, and the Colorado River District. If this agreement receives the endorsement of all the negotiating parties, Denver Water will have agreed to stop expanding its infrastructure on the Western Slope, except in the unlikely circumstance of the principal organizations of western Colorado supporting such an expansion: "Denver Water may develop any new water project on the West Slope only with the prior approval or under good faith consultation with the West Slope, depending on specified conditions."[41]

A second provision of the agreement gives Denver Water new options for responding to the dilemma of suburban areas like Douglas County. As reporter Allen Best summarized this, "A key component of the deal is giving Denver unchallenged authority to sell recycled water to communities in the metropolitan area's southern suburbs and exurbs, . . . which currently depend overwhelmingly on aquifer-tapping wells that are both costly and diminished in their deliveries." The important dimension to note here is Denver's provision to the suburbs of water that would be treated by a process sometimes spoken of as "from toilet to tap." And, returning to the idea of limitations on further diversion of water, the agreement carried abstention provisions that required Front Range "recipients of service from Denver Water [to] agree to abstain from seeking additional Colorado River water" in the headwaters above the confluence with the Gunnison River. In other words, if water providers in Douglas County signed on to receive reused water from Denver, they would also agree to forswear initiating transdivide diversions of their own from this portion of the Colorado River basin.[42]

And what did Denver Water gain from accepting the requirement to secure Western Slope consent to diversions? The agreement "moves forward an important project for the enlargement of the existing Gross Reservoir (the Moffat Project) which will provide additional water and enhance system reliability for Denver and its service area." In other words, the Western Slope participants in the negotiations had made their peace with the Moffat System expansion, putting a distance between themselves and the neighbors of Gross Reservoir.[43]

The agreement included provisions by which Denver Water would contribute $25 million as well as water from its Western Slope facilities. Much of the money would be used to support environmental mitigation and enhancement plans—that is, projects to reduce the environmental problems caused by removing even more water from the Fraser River (to be moved, via the Moffat Tunnel, across the Continental Divide, to storage in the expanded Gross Reservoir). Some of these measures would include redesigning the river channel to a more appropriate size for carrying the volume of water left in it after the diversions and augmenting that volume with water from Denver Water facilities during low-flow times of the year.[44]

Very soon after the agreement was announced, spokespeople for Trout Unlimited were expressing the conviction that Denver Water's commitments to pay for remedies for these problems are not yet proportionate to the scale of the problems. The increase in stream temperature, especially in midsummer, was a matter of particular concern, and Trout Unlimited's leaders continued to make a case for greater expenditures to maintain sufficient instream flow and to reconfigure the streambed in order to keep a depth of water in the river that would moderate temperature extremes.[45]

While a draft agreement had been arrived at and released to the public, ratification by all the parties was not a foregone conclusion. But Denver Water had participated fully in the negotiation, with Chips Barry playing a big part in its launching and his

The water that flows through the Moffat Tunnel has been diverted out of the headwaters of the Fraser River. Opponents of the proposal to raise Gross Dam fear that the additional diversions would overtax the river.

successor, Jim Lochhead, carrying it through to completion. In its concession that future projects on the other side of the Continental Divide would only go forward with the consent of agencies and organizations representing the Western Slope, *Denver Water in 2011 differed profoundly from Denver Water in 1990.* This was an organization whose leaders, on their own, now addressed the subject of limits, a message that no longer had to be blasted repeatedly into the zone of their attention by lawsuits. Contrary to the portrait of water organizations as inflexibly conservative and unable to reassess their guiding values, Denver Water had acted on the declaration that the board had made in the IRP that "'business as usual' could no longer be usual," and the "old assumptions would have to be examined anew."[46]

The 2011 agreement did hold open the possibility that in the future, Western Slope residents might consent to another major diversion structure. And yet, as Best has written colorfully, many people in western Colorado continue to view the diversion of water across the Continental Divide "as moral thievery and political thuggery."[47] Of course, opinions in the future could take unexpected turns. Denver Water might propose projects designed for the joint use of the Western Slope, with both sides of the state sharing in the benefits. But even those terms of consultation and explicit and intentional power sharing made for a striking contrast with the first seven decades of Denver Water's operations.

Had Denver Water changed, and if so, was the change more than window-dressing? Or, to put this in another way, if longtime Denver Water lead counsel Glenn Saunders had lived into these disorienting times, what would he have said upon learning that Denver Water had pledged to support new instream flow and recreational water rights on the Colorado River above Grand Junction? What colorful remarks would he have fired off about the idea of Denver promising to dedicate senior water rights to Grand and Summit County in the Colorado River's headwaters? Even letting his imagination run wild, could Saunders have conjured up an era in which Denver Water would agree "to make a good faith effort to identify which of its West Slope conditional water rights might be needed and to abandon those conditional water rights that it deems are not needed"? Various novel terms and phrases in the Colorado River Cooperative Agreement provided their own measure of the change between the era of Glenn Saunders and the era of Chips Barry and Jim Lochhead: "use of reusable return flows," "accelerated conservation plan," "environmental enhancement fund," "measures to reduce nutrient loading," "Cooperative Effort process . . . for the purposes of improving aquatic habitat," and "Water Rights Peace Pact."[48]

While the tale of Denver Water's last two decades carries considerable human interest, it could only be a vexation to sensationalist writers and to directors, producers, and movie stars. Any movie audience expecting the hypnotic intrigue of the plot of *Chinatown* would soon be comatose with boredom as the camera tracked the last five years of intricate interplay of provisions, clauses, and qualifiers among bureaucrats seated around meeting tables. "Being the manager of a large municipal water utility," Chips Barry wrote a few weeks before his death, "is no way for an adult to make a living. It is to hold a mid-nineteenth-century water development and natural resource exploitation position while encased in an early twenty-first-century regulatory straightjacket tailored by federal, state, and local governments." And yet, as Barry knew, there was no shortage of adventure and adrenaline in this line of work. Compared to the "huge, overwhelming, and difficult" challenges of building new structures like Cheesman and Dillon Dams, it still required plenty of "creativity, intelligence, and persistence" to work "in a bureaucratic setting," negotiating agreements and dealing with environmental impact statements.[49]

Hollywood may not discover the actor who could make Chips Barry's story as compelling on-screen as it was in real life, but Barry himself took considerable pleasure in capitalizing on a providential similarity of moustache and physical build that allowed him to play Theodore Roosevelt for various audiences. Given how much the origins of Denver Water reflected and drew from the principles and assumptions of Roosevelt and his fellow Progressives, there was something uncanny on those occasions when a century melted away and Chips Barry, having reconfigured an agency that originated in Roosevelt's time, played the part and spoke the words of that most famous Progressive. The status quo has never looked more dynamic.

Although his leadership of Denver Water left no doubt that he was one of a kind, Chips Barry bore a striking physical similarity to Theodore Roosevelt, and on occasion he channeled the Progressive president in front of various audiences.

Lessons of Interconnection

Rural and urban places
Are tangled together lke laces.
They're like sister and brother;
They depend on each other.
They have *never* been opposite cases.

Conclusion

Turning Hindsight into Foresight

Denver Water as a Parable

Beyond the sundown is tomorrow's wisdom,
Today is going to be long long ago.

—Thomas Hornsby Ferril, "Here Is a Land Where Life Is Written in Water"[1]

The fatal blow to the conventional wisdom comes when the conventional ideas fail signally to deal with some contingency to which obsolescence has made them palpably inapplicable. This, sooner or later, must be the fate of ideas which have lost their relation to the world.

—John Kenneth Galbraith, *The Affluent Society*[2]

Every minute in every western city and suburb, the omnipresent lawn offers the promise of liberation from fatalism and from stale habits of mind.

Colorado's American settlers brought with them an intense loyalty and devotion to imported plants—grasses, trees, flowers, bushes. The passage of a century and a half has not dramatically reduced the conviction of the state's residents (on the Western Slope as much as on the Front Range) that bright green is the proper color for a summer landscape. In the twenty-first century, the color green receives almost universal deference as the color of nature. Even the most ardent environmentalists in the interior West refer to their cause as the green movement, despite living in a part of the planet in which bright green is usually the color of disturbance, resting on a major, purposeful redistribution of water to compensate for the unreliable rainfall. The environmental movement in the interior West might, in other words, be better labeled the tan or russet or, at the least, the olive-green movement.

Currently, 54 percent of the water that Denver Water provides

In a far cry from the dusty, young town that the *Rocky Mountain News* once described as "treeless, grassless, and bushless," today the Denver metro area's suburbs are amply shaded by trees, carpeted in bluegrass, and full of bushes.

to its customers goes to outdoor landscaping. The territory that the agency serves remains ornamented with plenty of bright green lawns, maintained with generous doses of water, fertilizer, and fossil fuel for mowers. The greater share of these lawns simply offer a visual display, rarely trod upon even for a pleasant round of croquet or badminton. For westerners coming to recognize the relative scarcity of water in the region, lawns seem to parade their pointlessness, making it hard to figure out why a utility trying to reduce waste continues to supply water to keep these landscapes so radiantly green.[3]

For many who question the need for the expansion of Gross Reservoir and the related diversion of Western Slope water through the Moffat System in order to fill that added space, the persuasiveness of the justification for outdoor irrigation in urban and suburban settings has already petered out. To people of this persuasion, the pleasant appearance of a lawn in an urban neighborhood is the departing station of a train of thought leading directly to a distant place where a reduced flow of water threatens the beauty and the biological richness of a once-abundant river. "It is economically impractical to supply water for all uses all of the time, given Colorado's semi-arid climate," Western Resource Advocates said in its statement on the Moffat System expansion. "Denver is located in a semi-arid environment that receives far too little rain to support Kentucky bluegrass without supplemental irrigation," the statement continues, making "the conversion to natural areas, whereby Denver Water replaces turf grass with natural grass and native flowers," into an admirable act of conservation.[4]

To water managers, however, lawns deliver a service that would be far from evident to most observers. Lawns are devices that receive water that would otherwise depart Denver unused. Besides stymieing any schemes that Nebraskans might have to benefit from the Western Slope importations that Denver Water has achieved at considerable expense, lawns offer a cushion if severe drought should arise. Without that cushion, demand would be hardened; take out the lawns, and water would be directed only to needs that would not be susceptible to restriction. You can tell people to quit watering their lawns, and they will engage in a moderate amount of grousing. But if you were to tell people to shorten their showers and hold off on washing their clothes, they will grouse up a storm (and not the kind that produces precipitation). If you were to go all out with conservation and ensure that homeowners used water only for necessities and scattered not a drop on bluegrass or marigolds, you would be in a nearly impossible position when a serious drought hits. Without the margin represented by lawns, reducing waste would mean pushing beyond the reduction of luxury, cutting painfully into necessity. Thus, to Chips Barry, former manager of the Denver Water Department, the lawn looked a lot like a dispersed version of a reservoir, holding the water that could, in urgent circumstances, be shifted to respond to genuine need.

And yet the more important arena for thinking about the future of the lawn may well be not the calculations of the water managers but the hearts and minds of western residents. For the great majority of urban residents, lawns look like pockets of nature within the city, rather than insults to nature and affronts to the planet. They carry associations that are the opposite of environmental disturbance and manipulation. In homes surrounded by smooth green grass, families can find refuges embraced in expanses of nature; in well-watered parks and playing fields, people of limited means can enjoy outdoor leisure and exertion that would otherwise require an unworkable investment of time and travel.

The lawn, as writer Elizabeth Kohlbert writes, "has no productive value. The only work it does is cultural." And, since cultural values are dynamic, the lawn is subject to the same sort of changeability that, for instance, caused the beaver hat and the extravagantly feathered millinery to plummet in popularity, much to the relief and pleasure of North America's beavers and snowy egrets. "This is perhaps the final stage of the American lawn," Kohlbert speculates. Planting and maintaining a lawn has, in her judgment, become a matter of habit and expedience: "We no longer choose to keep lawns," she says; "we just keep on keeping them." Individually and collectively, minds may lose traction when searching for the answer to the question, Just why is that we devote so much time, land, water, money, and energy to the well-being of lawns? As historian Ted Steinberg gracefully phrased this, "The lawn has entered a vulnerable period in its history."[5]

Much stranger transformations than a decline in affection for lawns have, after all, happened in American perceptions of nature. Following the twists and turns of a story over two centuries has the pleasant effect of reviving one's sense of contingency, improbability, and even implausibility. The experience deepens one's humility when it comes to confident predictions about the future.

The well-watered lawn may seem to hold an unshakable place in American loyalties and preferences. And then again, trends and priorities may swerve and zigzag around, and green grass may, in a generation or two, be reconstituted as an inexplicable enthusiasm of the people of the past, who put unimaginable effort and exertion into raising a crop that no one would buy.

It is rarely, in other words, a rewarding practice to take customs for granted, to steer by unexamined assumptions, and to defer to what John Kenneth Galbraith called "the conventional wisdom." The history of Denver Water provides us with prime examples of, to use Galbraith's phrasing, "the fate of ideas which have lost their relation to the world."* The following pages apply the historical story presented in this book to the reconsideration of some of these well-entrenched ideas. I know they are well-entrenched because, until recently, I believed them myself.

The Causal—or Is That *Casual*?—Connection between Water and Growth

Mistaken Assumption Number 1: *The supply of water and the rate of population growth and residential development are inherently and inevitably intertwined. To increase population growth and residential land development, add water. To limit population growth and residential land development, stop adding water. Thus, agencies like Denver Water could control growth if their leaders would face up to their responsibilities.*

Neither prophets of doom nor prophets of progress get a lot of cooperation from the future. In the 1980s, Denver Water's leadership predicted that growth would screech to a halt and the economy of Denver would falter if the Two Forks Dam were not built. "Rejection of Two Forks will have a devastating impact on the Denver metropolitan area," Monte Pascoe of the Denver Water Board predicted.[6] The next years mocked his every word. Two Forks was not built. A carnival of growth in the Denver metropolitan region took place.

For its first seven decades, Denver Water's leadership alternated between proclamations of progress and predictions of gloom. After a round of celebrating its crucial contributions to local prosperity,

* While I am adopting Galbraith's concept, for the sake of clarity I am substituting the term *mistaken assumption* for *conventional wisdom*. I sum up each section with what I believe to be a better assumption, a better place to start one's thinking about water.

the agency would declare that if it did not receive permission and support for its next project, the well-being of Denver would head straight for a cliff, and landing at the bottom of that cliff would mean certain civic death. These grim prophecies usually came with a precise deadline: the year when the projected demand for water would exceed the current supply. If Denver Water were not allowed to increase the water supply on the Front Range, growth and prosperity would end.

No foresighted enhancement of the available water?

Growth screeches to a stop.

Denver Water and its opponents have, on occasion, united in the assumption that water and growth are causally connected in the most direct way. Coloradans who supported growth and Coloradans who opposed have converged on the belief that water is *the* factor that determines whether growth escalates, continues, sputters, declines, stops, or revives. Believers have stated this conviction as if they spoke of a natural law. Writing a short history of the department in 1944, longtime employee Walter Eha placed this belief at the foundation of his thinking. "Every man of vision living in this dry climate, who has looked back upon history and forward to the future, has understood that Denver's continued growth was and is absolutely dependent upon the development of her water supply," Eha declared. "The development of water resources has been the key to the development of the Denver area." Three decades later, holding an exactly opposite opinion of the desirability of growth, Alan Merson, the regional administrator of the Environmental Protection Agency, nonetheless matched Eha's belief in the causal connection between water and growth: "The record is very clear," Merson said in 1977, that "the availability of water determines whether growth occurs, where it occurs and the rate at which it can occur." Plenty of other knowledgeable folks have sung in this chorus. "Economic growth and development in the Denver metropolitan area depend on the availability of water," scholar Brian Ellison wrote. By the mid-twentieth century, "suburban leaders realized, like Denver's early progressives, that water was the key to controlling and benefitting from growth."[7]

In the early nineteenth century, starting from different premises and using very different language, explorers Zebulon Pike and Stephen Long had observed the scarcity of water on the Front Range and declared that this factor would constrain and even prohibit

American settlement. Two hundred years later, a variation on Pike and Long's idea has taken the form of a faith in the concept of carrying capacity, by which the maximum number of human beings, like the maximum number of wild animals or plants, would be determined by the natural environment. More than any other resource, water is cast as nature's enforcer. At an identifiable moment, when the numbers of people come to exceed the water supply, growth is supposed to hit a nonnegotiable limit with the force of a car full of crash dummies slamming into a very sturdy wall.

The problem with the idea of carrying capacity arises from the fact that human beings are a very peculiar species. Mule deer and prairie dogs do not get degrees in engineering, and their capacity to rearrange and reconfigure the location of the resources they need is correspondingly tiny. And, in an equally consequential difference, Americans colonizing western locales placed a big cushion in their water use. Their biological needs for water are small, but their social and cultural desires for water are vast. They use water for lawns, golf courses, swimming pools, fountains, and water parks. They would not die and their population would not crash if they were to find themselves forced to cut back in these luxuries. In other words, in determining the limits that water might impose on a human settlement, while it is important to know the sheer numbers of the population, it is even more important to know how much water individuals in that population are using. To arrive at a number of any value in estimating how many people could be supported by the available water supply, you must multiply the number of people by the number of gallons of water they use. Even when Denver Water's leaders were enthusiastic doomsayers who proclaimed that an inability to secure more water would soon produce disaster, they did not—and could not—say that Denverites were at risk of collapsing on their withered lawns and golf courses, perishing from hunger in long lines at water-restricted car washes, diving cataclysmically into emptied swimming pools, and thereby dying off. Despite the omnipresence of an enormously popular rhetorical flourish, very few Coloradans—unlike millions of others in much tougher circumstances on the planet*—live in

* The United Nations reports that "a third of the world's population is enduring some form of water scarcity. One in every six human beings has no access to clean water within a kilometer of their homes. Half of all people in developing countries have no access to proper sanitation" (United Nations, "The Global Water Crisis," online at www.un.org/works/sub2

circumstances where water is a matter of life and death.

Water, moreover, is far from the only factor that makes population growth possible. J. Gordon Milliken stated this proposition with great clarity:

> How large a role does water play in development? In a semiarid region, a reasonably secure supply of potable water is no doubt a major prerequisite of land development. Residential, commercial, and industrial land uses all require water. But the availability of water is not the sole factor influencing the pace and pattern of urban development.

Driving his point home, Milliken then offered a weighty list of additional factors governing the pace of growth:

- Cost and location of raw land;
- suitability of available land (in terms of slope, drainage, and characteristics, etc.);
- local government development policies and approval requirements;
- availability of other utilities (for example, electricity and natural gas) and services;
- transportation access;
- social and physical amenities (site-specific and regional);
- demand conditions in the local market (influenced by population growth and other demographic factors); and
- local, regional, and national economic factors (for example, interest rates).[8]

Repeatedly in the history of Denver Water, people acted on the assumption that increasing the water supply necessarily and directly increases the population. The border between belief and fact proved to be very permeable; strongly held convictions can shape human conduct and thereby can seem to confirm that the conviction was accurate. If, in other words, an influential group

.asp?lang=en&s=19). In his recent book, *Water: The Epic Struggle for Wealth, Power, and Civilization* (New York: Harper Collns, 2010), Steven Solomon makes a compelling argument for access to clean water as a basic human right, reminding us "how dehumanizing and economically crippling the lack of water for basic needs can be" (493). For a reckoning with problems of equity in water in western locales, see F. Lee Brown and Helen M. Ingram, *Water and Poverty in the Southwest* (Tucson: Univ.of Arizona Press, 1987).

believes that more water means more growth, they will *act* in ways that make this a self-fulfilling prophecy. They will increase the water supply, and when more people appear and pitch into the use of that water supply, the influential group will be certain that these two phenomena join together as cause and effect. Neither the people who increased the water supply nor the newcomers who took advantage of this will ask *how much water the new arrivals (and the old-timers who preceded them) actually needed*, nor will they stop to recognize the many other factors that played a part in this sequence.

With a different assumption about how much water people require, could the population grow substantially without an increase in the water supply? This is well within the reach of the possible. A lot more people could live with a lot less water *if* they recalculated the meaning of *need* and also if significantly higher prices invited them to think harder about the amount of water they used.

Many thoughtful and observant people have questioned the tightness and immediacy of the connection between increasing water and increasing growth. During the Two Forks fight, the Colorado Environmental Caucus considered "water an ineffective growth control mechanism." As one leader of the caucus, hydrologist Daniel Luecke, put it recently:

> In looking at it both regionally and nationally . . . you can't stimulate [growth] by supplying water nor can you prevent it by constraining access to water, in part because there are always options and in part because water is not a very expensive commodity. You can [stimulate or prevent growth] with transportation; you can do it with institutions of higher education; you can do it with airports; you can do with social services; but you can't do it with water.

With characteristically blunt phrasing, environmental historian Hal Rothman offered a variation on the same point: "No American city has ever ceased to grow because of a lack of water."[9]

Human beings have reason to be very attracted by the idea of having nature make our decisions for us, communicating its decrees through rates of precipitation and acre-feet of streamflow. But decisions to limit growth are decisions human beings must make, despite the pleasure we take in imagining ways to remove that weighty burden from our own shoulders and to delegate it to nature itself.

The Denver Water Board has been remarkable for its success in increasing the supply of water for residents, but it has also been remarkable in its success in periodically asking for conservation and efficiency in the use of water. With early calls for reduced use in the drought years of the 1930s, the effort to define the Blue Line to restrict the domain of service, the adoption of metering, the pioneering campaign for Xeriscaping, and the creation of tiered rates, Denver Water has accumulated a track record of exploring alternatives to the cycle by which increased demand and increased supply seem to chase each other in an upward spiral.

One of the most interesting questions in the twenty-first-century West is this: How much power do agencies like Denver Water have to direct the behavior of their citizens, clients, and consumers toward a sustainable relationship with water? Can an organization of this sort require citizens to distinguish their desires and whims from their actual needs and then to act on those distinctions, insisting on their needs and yielding on their desires and whims? Can an agency like Denver Water, well-positioned by virtue of having looked ahead and secured water rights early on, design and follow a policy of tough love in urging its neighboring communities to match their ambitions and plans to their actual capacities and assets?

The answer to these questions seems to be maybe not.

Denver Water's charter gives the agency significant powers and responsibilities. The Board of Water Commissioners has "complete charge and control of a water works system and plant for supplying the City and County of Denver and its inhabitants with water for all uses and purposes." In responding to neighboring communities, the charter adopted in 1959 specifies that "The Board shall have power to lease water and water rights for use outside the territorial limits of the City and County of Denver, but such leases shall provide for limitations of delivery of water to whatever extent may be necessary to enable the Board to provide an adequate supply of water to the people of Denver." The original charter and the amended charter, in other words, have from the beginning been unambiguous that the board's first obligation is to the residents of the City and County of Denver, not to the residents of its suburbs, of other parts of the state or of the Colorado River Basin.[10]

Conspicuously absent from this charter is even a hint of the power that would enable the water agency to determine how citizens are to live in an arid region. The charter delegates no power to tell residents where to live or which lands to develop for homes and which to leave undeveloped. If it remains in compliance with its charter, Denver Water cannot wield the authority to encourage or discourage population movement into the area. The right of Americans to move and to choose their places of residence has not been placed under the agency's jurisdiction. And it is an interesting and lasting paradox that key attractions, drawing in new residents to the Denver area, are the many days of bright sunshine and the escape from the discomforts of eastern humidity, exactly the features of the climate that constrain the natural water supply.

On July 8, 1972, when *The Denver Post* printed a letter from a professor at the University of Colorado, the public expression of wisdom on the actual extent of the power that Denver Water could assert over growth reached its high point.[11] Charles Howe began with a good-natured recognition of the achievements of the Denver Water Department, "an expert organization for the provision of water." "They know their engineering," he wrote, "and have a great system to prove it." An economist who was no slouch as a historian, Howe provided a capsule summary of the agency's first half century of existence: "The board is caught in the age-old dilemma of most water authorities of being charged with supplying water while having no control over major determinants of the demand for water, nor having any clear-cut guidelines as to what those determinants are likely to be in the future." The public, Howe noted, had a limited emotional repertoire when it came to attitudes toward the agency: indifference when the "supply is adequately provided" and "wrath . . . if water shortages should occur." Anticipating that wrath, Denver Water had understood getting control over water while it is available to be its priority.

While Howe believed that the board could and should put more effort into presenting alternatives to the public, other forms of criticism struck him as unfair. His assertions deserve quoting at length:

> The board should not, however, be blamed for the gross failure of the governor and legislature to provide meaningful guidelines for land use planning at the state level. To flail the water board for failing to control growth is really beating a straw man. If unwarranted, unplanned growth destroys the beauty of Colorado, it won't be because of the water board; it will follow from the absence of meaningful land use planning.

In the most forthright way, Howe reminded his readers that Denver Water had no authority to overrule the choices of people who wanted to move to this attractive region nor to prohibit the reallocation of land from farms to suburbs. Condemning the agency for failing to exercise power that it did not possess could never be more than a project in ineffective lamentation.

One would give a great deal to know what was in the mind of Denver Water counsel Glenn Saunders when he clipped Howe's letter from the newspaper and pasted it into his personal scrapbook. Even if we cannot know much beyond the fact that Saunders thought that Howe's commentary was worth preserving, the documentary evidence of Saunders's knowledge of this letter offers an irresistible temptation to hindsight speculation. What if the Denver Water Board, at Saunders's suggestion, had seized this opportunity and placed Howe's spirited statement—"To flail the water board for failing to control growth is really beating a straw man"—at the top of their letterhead?* Just below the quotation, they could have placed the invitation, "Let's talk about this!" What if Howe's sharp commentary on who actually held authority over growth had become a frequently cited point of reference in hearings, courtrooms, federal offices, and in rooms where environmental activists met and planned?

Contrary to this flight of imagination, Howe's letter seems to have carried little impact in its own time. Energetic flailing of Denver Water for failing to control growth continued without interruption after the publication of Howe's letter. And yet, if few at the time paid attention to Howe's letter, his statement carries undiminished relevance four decades later. "The Water Board is not a social engineering body," he wrote. "They have no expertise in planning the broad aspects of the good life in Colorado." At the risk of crossing the boundary between the religious and the secular, that remark qualifies as God's truth.

---

* If this idea seems preposterous, it is important to remind readers that the local newspaper *The Denver Post* has led the world in the use of quotations in this manner. The weather page carries the statement "'Tis a privilege to live in Colorado," while the editorial page carries the somewhat mystifying aphorism "There is no hope for the satisfied man," with both quotations originating with the *Post*'s founder, Frederick Bonfils. In that context, placing on stationery the aphorism "To flail the water board for failing to control growth is really beating a straw man" actually fits in regional practice.

Better Assumption Number 1: *Water is only one factor in population growth and not always the most important one. Controlling water does not necessarily translate into authority over growth.*

The Wicked Ways and Centralized Power of the Hydraulic Empire

Mistaken Assumption Number 2: *Power over water in the American West has been concentrated in the hands of a small, centralized, somewhat toxic elite.*

Through much of the twentieth century, representatives of the Western Slope and of Front Range suburbs characterized Denver Water as an autocratic, authoritarian, and imperial power. Much of this book shows that characterization to be exaggerated, even though it is perfectly understandable why rivals and opponents saw Denver Water in those terms. Despite condemnations of its vast and unconstrained power, Denver Water maneuvered its way through a world densely populated by competing agencies and entities, ranging from suburban governments to the US Supreme Court, from the US Bureau of Reclamation to the Colorado Supreme Court, from the Colorado River Water Conservation District to the US Army Corps of Engineers. Operating in this crowded arena, Denver Water never approached a position of unilateral power over water in the state of Colorado.

With this statement, I part ways from an older orthodoxy in the interpretation of western water history and join, instead, in the assertion of the fragmentation and decentralization of authority over water. As American attitudes toward nature changed directions in the last half of the twentieth century, the publication of two very influential books, Donald Worster's *Rivers of Empire* in 1985 and Marc Reisner's *Cadillac Desert* in 1986, reflected that shift and established a new orthodoxy in the field. To both Worster and Reisner, the history of the development of water in the West followed a plot (in both senses of the word) that led to the centralization of power in the hands of a small, inflexible, undemocratic, entirely self-interested elite.* "The hydraulic society of the West," Worster asserted in a widely quoted statement, "is increasingly a coercive,

* Marc Reisner kept readers' attention focused on the arbitrary powers of the Bureau of Reclamation and the US Army Corps of Engineers, and yet in a number of passages, he acknowledged that the undertakings of those agencies enjoyed widespread popular support.

monolithic, and hierarchical system, ruled by a power elite based on the ownership of capital and expertise."[12]

When shifting paradigms, a student of mine once noted that it is important to remember to put in the clutch. In the last two decades, other scholars have doggedly deployed such a clutch to shift the paradigms for interpreting western water history. Western historians Donald Pisani and Norris Hundley have called into question the ostensible consolidation of power over water in the hands of a coordinated and even conspiratorial few, putting forward stories structured much more by contesting groups than by a power-hoarding elite. In his most effective challenge to the image of a coherent and cohesive empire of control over water, Pisani makes a compelling point about the fragmentation that characterized the efforts of westerners to claim and develop water. "Fragmentation," Pisani writes, "resulted from a pervasive mercantilism that pitted community against community and state against state, from intense competition between regions within the West, . . . and from a decentralized system of government." Water policy thus arose from a political and cultural atmosphere of "persistent suspicions and irreconcilable differences," an inauspicious foundation for the imposition of monolithic power.[13]

In the development of water in California, Hundley acknowledges "the appearance of a new kind of social imperialist whose goal was to acquire the water of others and grow at their expense." And yet, while noting the "monumental conflicts and social costs" created by these energetic "social imperialists," Hundley declares that "at the same time, this is a story of extraordinary feats of fulfilling basic social needs when communities mobilized and focused their political energies on providing abundant clean water to multitudes of people who clearly wished that to be done." Rather than a centralized process purposefully orchestrated and conducted by a narrowly self-interested elite, Hundley "describes the activities of a wider and often confused and cross-cutting range of interest groups and bureaucrats, both public and private, who accomplish what they do as a result of shifting alliances and despite frequent disputes among themselves."[14]

For advocates of the earlier paradigm that Pisani and Hundley have challenged, and certainly for Marc Reisner in *Cadillac Desert*, the prime example of centralized, coercive, and secretive power at work appeared in the case study of the city of Los Angeles's acquisition of water from the Owens River Valley. The influential movie *Chinatown* was the cinematic apogee of this school of historical judgment. But even this widely known tale has recently submitted to convincing revision. In 2006, Steven Erie pointed out major flaws in the conventional wisdom that structures the far-better-known version of this tale. Rather than the *Chinatown* version, in which "unscrupulous developers in league with conniving water officials secretly orchestrated water megaprojects for private financial gain," Erie declared that it was time to "retell much of the region's twentieth-century water and development saga as an innovative public venture rather than a sordid secretive affair." The notorious engineer for Los Angeles, William Mulholland, Erie argues, did not pursue personal profit, and even if individual land developers in the San Fernando Valley later profited from the delivery of water from the Owens Valley, they were not themselves the initiators or arrangers of the original plan for the diversion. Erie's work has failed to inspire filmmakers to produce an answer to *Chinatown* that would feature this factually accurate, comparatively colorless version of the Owens Valley tale.[15] Given the pleasure that Americans seem to find in scaring themselves silly, don't expect Hollywood to race to seize this opportunity.

Better Assumption Number 2: *The acquisition, development, allocation, and management of western water have been processes characterized by fragmentation and competition as much as (if not more than) the exertion of centralized power.*

Rural Virtue and Urban Wickedness: The Long-Awaited Retirement of Thomas Jefferson, Agrarian Dreamer and Urban Condemner

Mistaken Assumption Number 3: *In opinions on and judgments of competing demands for water, use for farms and ranches carries a greater ethical integrity and is more justifiable than the use of water for environmentally parasitic cities and suburbs.*

As Steven Erie's fresh perspective on the Owens Valley story reveals, an unexamined but very influential assumption of rural virtue and urban wickedness has shaped common understandings of the history of urban water development in the West. In a lively passage worth quoting at length, leading historian of the urban West, Carl Abbott, joins with Erie and Hundley in sending the conventional wisdom on the Owens Valley diversion through a spirited reappraisal. Substituting *Denver* for *Los Angeles* and *Western Slope* for *Owens Valley*, readers will note, does not require much change in the contents of this passage.

> There are, of course, some problems with the [usual] moral understanding [of this historical episode]. . . . Los Angeles may have been shrewd in buying water rights, but it purchased something that Owens Valley residents had themselves transformed into a commodity. Owens Valley people had already manipulated the water through irrigation systems, so Los Angeles did not acquire and pervert something that was purely "natural." Left unanswered is the question of how far water can legitimately be diverted: one mile? ten miles? 200 miles?. . . There is a strong assumption that it is "unnatural" to live in Los Angeles, but is it any more natural to impose Euro-American agriculture on a semi-arid landscape only a mountain range away from Death Valley?[16]

As Abbott's energetic critique shows, any appraisal that instantly discounts and discredits the legitimacy of urban water development reveals its dependence on arbitrary and doubtful assumptions.

While readers of this book are encouraged to reach their own appraisals (and to convey them to the author), my own habits of thought have changed to match the perspectives of Pisani, Hundley, Erie, and Abbott. Before this recent conversion, with hundreds of students sitting in lecture halls mistaking me, sadly enough, for a knowledgeable authority, I followed the old version of urban condemnation when I told them the history of Los Angeles and the Owens Valley and of San Francisco and Hetch Hetchy. Looking back at my embrace of this orthodoxy, I am impressed by the degree to which I and many others of my generational cohort steered by the tenets of agrarian idealism when appraising the allocation of water in the West.

We know who led us down this path: President Thomas Jefferson took quotability to great heights in his characterizations of rural virtue and urban wickedness. "Those who labor in the earth are the chosen people of God," Jefferson declaimed, "if ever He had a chosen people, whose breast He has made His peculiar deposit for substantial and genuine virtue." Meanwhile, things did not look good for the cities, where workers depended on wages for the purchase of their subsistence: "Dependence," Jefferson said, "begets subservience and venality, suffocates the germs of virtue, and prepares fit tools for the designs of ambition." And then Jefferson ascended to the peak of antiurban rhetoric: "The mobs of great cities add just so much to the support of pure government, as sores do to the strength of the human body." As a "canker which soon eats to the hearts of [a republic's] laws and constitution," the American city had a few strikes against it after the most articulate of the nation's Founding Fathers characterized it as a symptom of a repellent disease.[17]

Belief in the inherent superiority of the enterprise of farming played a big role in shaping the material reality of the West. The passage of the Newlands Reclamation Act in 1902, directing the federal government to use public resources to expand irrigated farming in the West, demonstrates the power of the agrarian ideal. Down to our own times, the proportion of water directed to agriculture remains above 80 percent in western states. The Jeffersonian agrarian ideal has not only shaped the writings and teachings of academic historians; it has been a fundamental force in reshaping the western landscape.[18]

On the even more consequential front of popular culture, the rural areas have also trounced the urban areas. The mythic appeal of the West has simply been better situated, and shown to better advantage, in open space. Cowboys riding freely on handsome horses toward a distant horizon have been targets of admiration and objects of envy. There was something considerably less appealing in the circumstances of cowboys walking down dusty town streets and into claustrophobic saloons, tearing their hair (not an easy achievement with the regulation cowboy hat) over their obligation to protect a bunch of townspeople, usually rendered weak and witless and thereby not particularly fun to protect when they were caught in crossfire between bad guys and good guys. Countless movies and novels thus added vast force to Jefferson's antiurbanist cause.

Advocates for the preservation of nature made their own lasting contribution to the allocation of virtue to the countryside and wickedness to the city. For the better part of a century and a half, people who wanted to have nature left undisturbed and wildlife protected from human encroachment have not volunteered as enthusiasts and cheerleaders for cities. And yet, with the characterization of cities as places with qualities that were quite the opposite of the beauty, appeal, and power to uplift the human soul delivered by open spaces, antiurban sentiment posed an ironic obstacle to the well-being of those open spaces. The popularizing of an ardent affection for nature carried the corollary that a life lived close to nature was a better, more grounded, and more uplifting life. The material outcome of that belief was the rejection of city life and a vast increase in the number of homes built with picture windows with views reminding the homeowners that they were indeed close to nature. The proliferation of suburban and exurban residences meant a disruption of open landscapes and wildlife habitat. As historian Kenneth Jackson pointedly reminds us, "The first necessary condition for the unusual residential dispersal of the American people is a national distrust of urban life and communal living."[19]

In fact, concentrating human populations in dense urban settings is an effective method—possibly *the* most effective method—for reducing the impacts of human settlement on the landscapes and rivers of many regions, including the West. In *Green Metropolis,* David Owen uses New York City as his example of the environmental advantages of urban density. He acknowledges the degree to which his convictions defy conventional wisdom: "Thinking of crowded cities as environmental role models requires a certain willing suspension of disbelief, because most of us have been accustomed to viewing urban centers as ecological calamities." But Owen declares "the apparent ecological innocuousness of widely dispersed populations—in leafy suburbs or seemingly natural exurban areas" to be "an illusion." "Spreading people thinly across the countryside," he observes, "doesn't reduce the damage they do to the environment. In fact, it increases the damage, while also making the problems they cause harder to see and to address." Owen's observations have enormous bearing on western America: "Wild landscapes are less often destroyed by people who despise wild landscapes than by people who love them, or think they do—by people who move to be near them, and then, when others follow, move again." In his most compelling point, Owen declares that "you create open spaces not by spreading people out but by moving them closer together."[20]

The environmental standing of the city has indeed been undergoing a makeover. As environmental historian Martin Melosi wrote in 2000, "cities have the capacity, when properly designed, to use resources more efficiently than highly decentralized populations." In an opinion piece in *The New York Times,* University of Michigan urban planning professor Christopher B. Leinberger declares that "the good news is that there is great pent-up demand for walkable, centrally located neighborhoods in cities like Portland, Denver, Philadelphia, and Chattanooga, Tenn." "It is time," he asserts, to "build what the market wants: mixed-income, walkable cities and suburbs that will support the knowledge economy, promote environmental sustainability, and create jobs."[21]

What If Henry David Thoreau had written more about the charms of Boston than the appeal of Walden, if John Muir had featured the attractions of San Francisco over the beauty of the Sierras (though this might have posed some trouble for his crusading against the Hetch Hetchy Dam), if Edward Abbey had written more about good times in Hoboken, New Jersey, than about the entrancing qualities of the Four Corners canyonlands? The persuasive strategies of preservationists and environmentalists would have been dramatically better aligned with their goals. A society that felt better about urban density would be a society better positioned to support and sustain healthy ecosystems in the great outdoors.

And yet support for the transfer of water from agriculture to cities is by no means the obvious conclusion from this line of reflection. The trend toward agricultural water transfers, in which urban and suburban areas purchase rights from farmers and ranchers, is shaping up as a major force affecting the rural landscape that, for so many, represents the "real West." In the twenty-first century, the habit of mind that pits rural interests against urban interests proves to be distinctly un-useful. In down-to-earth reality, urban well-being and rural well-being are more intertwined than they are distinct or reciprocally injurious. A thriving rural world is an asset for a neighboring city; the proximity to open spaces is, after all, a principal reason why people want to live in a city like Denver. Ranching and farming are essential forces in the preservation of the West's open horizons. Moreover, a growing enthusiasm for local food production offers an expanding market for farmers in the vicinity of cities, adding another argument for pursuing the hope of a shared urban and rural prosperity.

A recent report from three environmental groups, "Filling the Gap: Commonsense Solutions for Meeting Front Range

Water Needs," puts forward a "Smart Principle" that connects very directly to this line of thought, offering this advice to decision makers: "Recognizing that market forces now drive water reallocation from agriculture to municipal uses, structure such transfers, where possible, to maintain agriculture and in all cases to mitigate the adverse impacts to rural communities from these transfers." To put that in language a little less pitched to professionals working in the field of water management: "Don't waste effort on fighting an unmistakable trend of moving water, by sales and leases, from farms and ranches to cities and suburbs. But do seek out every opportunity to keep farms and ranches from going out of business, and at the very least, when the supplying of water to a city will produce injury to a rural area, find a way of repairing or reducing that injury." In practical terms, this recommendation means stopping these transactions short of permanent, irreversible sale of water rights: "Innovative arrangements, such as rotational fallowing, interruptible supply agreements, water banks, crop changes, and deficit irrigation,* could allow for temporary transfer of irrigation water to municipal uses without permanently drying irrigated lands." And, with more attention to conservation and efficiency, the demand for water in irrigated farming could be significantly lowered, leaving more water available to build an alliance between rural areas and urban areas.[22]

Exploring the history of Chicago and its hinterland, historian William Cronon drew conclusions that illuminate and deepen the history of water in Colorado. Cronon initially thought he saw "reassuringly sharp boundaries between city and county." But "the moment [he] tried to trace those boundaries back into history, they began to dissolve." Urban and rural landscapes, he came to believe, "are not two places but one. They created each other, they transformed each other's environments and economies, and they now depend on each other for their survival. To see them separately is to misunderstand where they came from and where they might go in the future."[23]

Better Assumption Number 3: *There are many good reasons to reject old appraisals of the distribution of virtue and the corresponding allocation of*

* Deficit irrigation is a system of reducing the water applied to crops in a strategically timed manner, when the water stress will have minimal impact on the plants' productivity.

water between rural and urban areas and to search instead for the ties that link the well-being of both domains.

An Awkward Debt, with Interest Accumulating: Society, Engineers, and Infrastructure

Mistaken Assumption Number 4: *Although members of the engineering profession should be reprimanded when they make messes, they should generally be left to their own devices. Infrastructure works best when the great majority of citizens never have to pay any attention to it.*

The history of Denver Water provides a rich case study in the force of human ingenuity, particularly as that force has powered the profession of engineering. The plans and actions of engineers, more than any other factor, produced the transformation of the Denver metropolitan region. And yet, in a curious pattern in the writing of western American history, other occupations have drawn the lion's share of attention, and engineering has been left waiting for recognition. The historical field has long been well supplied with studies of explorers, fur trappers, cowboys, ranchers, farmers, and loggers. Studies of engineers are very rare, even though engineers have been central to the key enterprises in western development: mining, railroading, water diversion, road building, energy production, and urban development.*

A heightened attention to the profession of engineering is of practical, as well as scholarly, consequence because the relationship between the American people and the engineering profession is in need of some significant reengineering itself. To use the phrasing of historian Martin Melosi, engineers have been "facing a major transition in how they were perceived and how they perceived their function." For more than a century, citizens, customers, clients, and public officials have said to engineers, "Get us what we want in water supply; make sure that the water is clean and cheap; and be careful not to place any unsightly disruption in the landscapes that we love." In the history of Denver, thousands of people lived lives of considerable comfort under these less-than-fair terms. Reckoning

---

* The conspicuous exception has been the attention paid to the engineers who surveyed the routes and supervised the construction of transcontinental railroads, though even in that case the attention has been more on colorful, self-dramatizing individuals like Grenville Dodge than on the broader significance of the profession.

with that history gives us a prime opportunity to note—and act on—the need for better terms for this relationship. A major project for the twenty-first century rests on persuading those who benefit from vast engineering projects to forge a more honest assessment of the consequences of their own needs and to face up to their complicity in—and their many benefits from—the work of engineers. As writer Diane Ward put it, "for the most part engineers have been only as good—or as bad—as the tasks they were given by the rest of us."[24]

Also posing an obstacle to the improvement in the working relationship between society and the engineering profession is the quick impulse to condemn technology. Engineers have operated with insensitivity toward and contempt for nature, the complaint goes. They have been unrestrained in the exercise of their power. They put all their faith in technological fixes and never raise the possibility of wise adaptation to natural constraints. This line of thought presents two problems. First, to deal with the consequences of a century and a half of rearranging the natural environment, we are very much dependent on the skills, insights, and good will of engineers. The need for their ingenuity shows no signs of shrinking. And, second, the litany of condemnation and complaint ignores a spirit-lifting development of recent time: the rise of the profession of environmental engineering, in which engineers, with impressive clarity and honesty, assess the dilemmas that their predecessors have created and seek to correct or reduce the injury, damage, or loss produced by previous professional practices. Legal scholar Robert Glennon's observation is certainly accurate: "In the past, when we needed more water, we engineered our way out of the problem by diverting rivers, building dams, or drilling water. Today, with few exceptions, these options are not viable solutions."[25] The solutions of the present and future will, however, still require engineering our way—but guided by the values of the emerging practices of the environmental engineers.

As our shared inheritance, the vast American infrastructure built in an increasingly distant past is showing its age. Infrastructure briefly rockets to national attention when it breaks down in a sudden calamity, while matters of maintenance and replacement languish from inattention, resistance to public expenditures, and dislike of big government. Indeed, without the stimulus of crisis, the very word *infrastructure* currently works *not* to ignite enthusiasm and recruit support but to send minds wandering in search of something more interesting to contemplate. As the history of Denver Water shows, infrastructure enjoys the briefest of glory days: at the moment of their completion, dams, reservoirs, tunnels, treatment plants, and pumping stations will be saluted with a ceremony of completion, with a few solemn speeches and some rarely-very-spirit-lifting photographs of dignitaries engaged in ribbon cutting. When the ceremony is over and the applause dies down, the infrastructure is supposed to recede to a state of invisibility, delivering water in the manner of an unobtrusive, uncomplaining, self-effacing servant. The people who manage the infrastructure are supposed to keep an equally low profile, maintaining the infrastructure in a manner that keeps costs low, prohibits interruptions or breakdowns in service, and somehow holds off time's legendary aggressiveness as an agent of demolition, relentlessly deploying the tools of corrosion, abrasion, and gravity.

And yet, in the early twenty-first century, a cadre of brave people refuse to defer to the popular preference for keeping infrastructure out of sight and out of mind. A plucky group of public officials and pundits keeps reminding us that our well-being rests on a vast network of aging structures and facilities, while a widely shared resistance to paying the price for the restoration and rehabilitation of this network leaves our circumstances ever more precarious. When it comes to restoring and rehabilitating our sense of responsibility for and indebtedness to the systems that supply us with essential services, a visit back to the point of origin of these systems has unquestionable value. For the open-minded, there is also wonder and surprise in the recognition that, in a distant time, human beings so rapidly conceived of and built structures of such scale—*and such consequence.*

Better Assumption Number 4: *To help in the crucial cause of building a direct and honest relationship with technology and its creators, citizens benefitting from the water infrastructure should cultivate both the company of engineers and a livelier sense of personal responsibility. Engineers will be essential participants in finding solutions for the dilemmas generated by history, making hindsight condemnation of the profession into an unrewarding and even counterproductive sport.*

"Whiskey Is for Drinking and Water Is for Fighting Over": Mark Twain's Doubtful Career Achievement as a Water Policy Analyst

Mistaken Assumption Number 5: *Water is fated to produce conflicts, contests, and even wars because it is so important to every enterprise and undertaking and to human life.*

People who attend a water conference are almost guaranteed to hear at least one speaker (if not five or ten) quote Mark's Twain's legendary remark, "Whiskey is for drinking and water is for fighting over." Unending wear and tear from rubbing up against too many podiums and microphones has made this remark into the equivalent, in freshness and originality, of the story about the waiter who when asked by a customer what a fly was doing in her soup responded, "The backstroke."

Besides overuse, this old saying has two flaws: inauthenticity and inaccuracy. First, there is no evidence to support the assertion that Mark Twain coined this tedious aphorism.[26] Second, there is a good likelihood that the statement is only half-true. Whiskey *is* for drinking. Water, by contrast, is for every activity under the sun. Every now and then, it is for fighting, but it is also for negotiating, deal making, marketing, collaborating, and conserving.

Several years ago, publishers persuaded a sharp and thoughtful writer named Wendy Barnaby to write a book on water wars. She took this assignment willingly. Experts had gone to town with cataclysmic forecasts in the 1990s; Barnaby cites the particular example of "former World Bank vice-president Ismail Serageldin's often-quoted 1995 prophecy that, although 'the wars of this century were fought over oil, the wars of the next century will be fought over water.'" But her work on the book did not proceed smoothly. "Power struggles and politics," Barnaby came to realize, "have led to overt and institutionalized conflict over water—but no armed conflict, as there is over borders and statehood." Rather than going to war, countries far more often "solve their water shortages through trade and international agreements." The insights of a British social scientist, Tony Allan, showed how this worked. When a city or a nation imports food from rural areas, this adds up to a transfer of embedded water, or, in a term Allan increasingly used, virtual water. Every orange, strawberry, kernel of wheat, potato, or cucumber comes into being with a significant deployment of water.* Water-abundant areas could grow those products, and water-short but prosperous countries could buy them. Allan focused his studies on conflict-rich areas of the Middle East where, he found, arid countries dealt with their shortage by importing grain and its embedded water.[27]

Rather than going to war, Barnaby came to believe, nations "solve their water shortages through trade and international agreements." This presented a challenge, not only to Mark Twain's endlessly quoted apocryphal assertion, but also, more immediately, to the author's relationship with her publishers. Allan's analysis "killed [her] book." She "offered to revise its thesis, but [her] publishers pointed out that predicting an absence of war over water would not sell."

Concluding a book on water by calling attention to the death of Barnaby's planned book may seem an ill-fated move for a hopeful author. But there are irresistible satisfactions to going out on this note. Writer Alexander Bell chimed in on the chorus with Allan and Barnaby: "Cooperation on water," he wrote, "is very resilient." Water may be of such crucial importance that feisty leaders of nations have to come to their senses and negotiate. "The truth is," Bell concludes, "that water may be just as effective at disarming man's war instincts as it is as starting the fight."[28] Allan, Barnaby, and Bell offer a vitally important challenge to an ancient and fatigued piece of conventional wisdom about water's inherent and inevitable power to pit individuals and communities against each other. Applying their thinking to the history of western water offers strenuous but valuable exercise to westerners in the twenty-first century, precisely the kind of exercise that reduces gloom and enhances hope.

Consider the fact that the Colorado Constitution, despite the powerful forces of antiurban thinking in the American past, gives domestic water use the highest priority in times of water shortage. Consider, as well, this thought-provoking pair of numbers: Denver Water uses slightly more than 2 percent of all of the treated and untreated water used in the state to support more than 25 percent of the state's population.† To a remarkable degree, we inherit legal

* Allan's point usefully complicates the familiar phrasing by which urban water use is distinguished from rural water use; the importing of food grown in rural areas into a city like Denver also entails the importing of embedded, or virtual, water.

† According to Denver Water, the agency uses about 265,000 acre-feet per year. According to the

principles and an infrastructure that support the density of settlement that offers hope for the prospect of clustering people in ways that reduce environmental disturbance and disruption.[29]

Our predecessors on the planet went tooth and nail at the development of resources, carrying on as if there were no tomorrow. But there turned out to be a tomorrow, and we are its occupants. There are unmistakable assets, as much as unmistakable liabilities, in our inheritance from the past. In the inventory of the estate left to us by our predecessors, there is no argument for fatalism.

Thus, Wendy Barnaby gets the last word:

Better Assumption Number 5: *"It would be great if we could unclog our stream of thought about the misleading notion of 'water wars.'"*[30]

Colorado Water Conservation Board's Statewide Water Supply Initiative–2010 report, municipal and industrial uses divert 1.16 million acre-feet per year and agriculture uses 4.8 million acre-feet per year. Assuming a typical irrigation efficiency of 50 percent, the total diversion for agriculture is approximately 9.6 million acre-feet. Thus, the total diversion of treated and untreated water for municipal, industrial, and agricultural uses is approximately 10.76 million acre-feet per year. Denver Water's use of 265,000 acre-feet represents slightly less than 2.5 percent of this total. The 2010 census pegged the state's population at 5,029,196, and Denver Water serves approximately 1.3 million customers, or just over a quarter of the population.

AFTERWORD

Two Decades at a Western Water Utility

Some Reflections, Observations, and Occasional Insights*

Chips Barry, Manager, Denver Water, 1991–2010

Being the manager of a large municipal water utility is no way for an adult to make a living. It is to hold a mid-nineteenth century water development and natural resource exploitation position while encased in an early twenty-first-century regulatory straight-jacket tailored by federal, state, and local governments. Every group with an interest in water, development, or the environment takes a keen interest in making certain that you make no mistakes. But there are a few lucid moments, crystalline experiences, and watershed events that define the truth beyond and lay bare the essence of it all. I have had many of these moments. They were all events of profound and brilliant failure, but string these glistening moments of defeat into a strand and you have the pearls of an administrative career. Let me describe a few of these moments and distill from them whatever wisdom lies therein.

1. OUR WATER FUTURE IS A LOT MORE UNCERTAIN THAN IT USED TO BE.

Planning has always been a critical but unseen component of any large water utility operation. As long as water comes out of a faucet when open, no one thinks about where it comes from or whether the supply is sufficient. But the utility has to think about it. Planning for water utilities has changed enormously in the last ten years and it will continue to change because of changes in demographics, economics, and because of global climate change. The paradigms for water planning are changing. As some recent pundit noted, "When changing paradigms, be sure to put in the clutch."

* A talk given at a meeting of the Colorado Yale Association at The Fort Restaurant in Morrison, Colorado, March 28, 2010

Chips Barry standing behind a waterwheel in 2003. In the last decade of the twentieth century, he became the face of Denver Water.

In the past, water planning was a comparatively simple exercise of matching the supply of water with the expected demand for it. To determine future demand, you examined the past five years of usage, assumed no change in the pattern or timing of use or in the weather, and multiplied that usage by a percentage (typically 1 or 2 percent) that reflected the population growth in your area. The formula on the supply side was similar—determine what your water rights would yield on the average or on a dry-year basis. If supplies were sufficient, you were fine; if not purchase some more water or build some new storage or both.

Although this planning methodology may still work in some locations, to many of us in the West it seems hopelessly outdated and naive. In the first place we no longer just "determine" demand; we make a very concerted effort to manage it. We have succeeded in reducing demand in Denver by about 20 percent in the last eight years and believe there is more that can be done without adversely affecting any customer's quality of life. But unmanaged demand is more easily predicted than managed demand. Plus, what effect will global climate change have on the demand for water? How effective will be our efforts to manage that demand? Figuring out how much customers actually will demand is a good deal more uncertain than it was in the past.

Determinations of future water supply are equally difficult. We know the climate is warming and we can probably predict with some accuracy the range of future ambient temperature in Colorado. The hard part is that no one can predict what happens to precipitation in the central Rocky Mountain area. The best guess now is that precipitation will be much more variable than before. As Yogi Berra once said, "The future just ain't what it used to be." The past hydrologic record was a superb indication of what the range of future precipitation would be. It is still the best indication, but it seems to be a lot less reliable than before.

For many, the answer to uncertainty is often to get more data, which allegedly will enable a better decision. However, with respect to both water supply and water demand, more data will not necessarily be of much help. The uncertainties will remain, and water planners must instead grapple with how much risk is tolerable and how much reliability is required. In the end, we are left with yet another Yogi Berra quote: "Forecasting is a difficult thing to do, especially if you are talking about the future."

2. Policies often have unintended consequences, and no good deed goes unpunished.

My example in this case comes from early years at the Denver Water Department where I was determined to put in place a new policy affecting all employees. The new doctrine was that I would maintain an open-door policy. The reaction to this pronouncement was quite something.*

The carpenter shop sent word back via the maintenance section that they were sure the door worked properly and if I had any complaints I should send them down through three layers of supervisors to the carpenter shop. Employees began to appear at the outer fringe of my office, but not to enter, only to peer in, and assure themselves that the door was in fact open. A few employees actually entered the office and announced that they were there pursuant to the open-door policy. However, the policy was so novel that they didn't know who was to speak or what business was to be conducted. Three supervisors asked whether they could station a monitor outside my office to find out who was utilizing the open-door policy. Rumors circulated that the manager was going to reduce staff by 30 percent, and that the first three hundred people through the open door would get the ax. Facilities management wanted to know whether I needed the carpet repaired because of all the traffic from the open-door policy, and the board wanted to know how I could get any work done if all I was doing was talking to employees. The real estate speculators from northern Douglas County viewed my open-door policy as an invitation to advocate that Denver provide them water from its "hoarded" supplies. The *Grand Junction Sentinel* ran an editorial suggesting that the open door should be used as an escape device for all the west slope water that Denver has been holding hostage for fifty years. The suburban proponents of Two Forks threatened suit if they were charged for the manager's time during open-door meetings but wanted copies of any notes of meetings with employees if Two Forks was the subject discussed. *Westword* suggested that reform was long overdue at the water department and these major policy shifts were certainly indication that change was under way. *The Denver Post* on the other hand opined that too rapid change led to a destabilized

* Readers should be advised that the story that follows is a parable or fable; while it conveys a certain truth, appraising its accuracy is emphatically not the point.—Ed.

institution, without direction, focus, or accountability. One local pundit linked the open-door policy to the forthcoming baseball season while another wondered in print if I was running for office.

As I have said, policies can lead to unintended consequences, and no good deed goes unpunished.

3. Colorado water law is almost always counter-intuitive, and at its worst it is Byzantine.

A few illustrations:

- In Colorado, we have tributary groundwater and nontributary groundwater. So far, fine, but we also have not-nontributary groundwater. Don't ask me what it is, because I can't explain it, but I am not making that up.
- Only in Colorado water courts can you file something officially entitled Statement of Opposition in Support.
- We have at least three kinds of water in Colorado other than tributary and nontributary. People will often refer to water with no adjective, but they will often clarify their meaning by saying "wet water" or "paper water." I don't think any other state specializes in paper water the way we do.
- In the words of art used in water law, you need to know that water diverted from the stream is not the same as water depleted, and that consumptive use is different from just plain use.

4. We have unfortunately politicized hydrology, and we will regret it.

In hydrologic terms, hydrology is becoming a tributary of politics. Politicizing something means using an otherwise noncontroversial event or occurrence to further a political agenda. I see that occurring in a variety of areas, and I am distressed by it. Dr. Kevorkian has politicized death, and the abortion and antiabortion forces continually challenge each other on the rapidly developing politics of abortion. Genetic research has now been politicized, and we now see the same tendency with respect to believers and nonbelievers of global climate change.

There is ample evidence that both El Niño and global warming are a fact. Both have dramatic short-term and long-term effects on climate, rainfall, and freshwater distribution. It is unclear whether there is a connection between global warming and El Niño occurrences. Does global warming cause El Niño? Does El Niño exacerbate global warming? Many think they know and see their political opportunity in advocacy of their answer. Both global warming and El Niño have occurred in the past. The critical question is whether human activity is making El Niño and/or global warming worse. Many believe that mankind is wholly or partially responsible for these events. To the extent that this allegation is believed, we will have succeeded in making hydrology a political issue. If the world believes that human activity causes El Niño with subsequent weather dislocations, storms, floods, droughts, mudslides, etc. and that human activity drives significant increases in the rate of global warming, then the efforts to alter human behavior in response to this occurrence will, without question, result in political pressure on the field of hydrology. Presumably there was a time when hydrology was pure science, when no political judgments were involved in determining how much water was flowing, how much groundwater was in reserve, or what the relationship was between the two. However, if politicians now believe they must have a stake in altering human behavior because of their concerns about climate, the questions about water availability, water flow, groundwater, etc. will be manipulated and used to serve an agenda that is either directed at limiting human activity or an agenda that is directed at allowing it to continue relatively unfettered. In this setting, hydrologic data is no longer neutral; it becomes a weapon for use by one side or another in the battle over the extent of allowable human activity. I see this in the debates about flow in both the Colorado and the South Platte Rivers. Conclusions about how much water is in the river, and how much water *should* be in the river, are coming to depend more on your belief about agriculture, development, and water diversion, than on hydrologic data. We've now reached the point where contending sides hire their own psychologists, accountants, and doctors, each of whom is expected to present testimony or a report supporting the point of view of the party who has paid the expert. I don't look forward to the day when we also have forensic hydrology and each side in an important water issue hires a hydrologist who will be expected to present a particular point of view that furthers the political agenda of the proponent. Yet, that is where we are going.

5. Water conservation is a surprisingly complicated subject, and it is not the answer to every water question.

For many, water conservation seems like a simple topic and an obvious good. Use less of the resource and you will not waste it. The more conservation you can do, the better off you are. Alas, it is not so simple. The first question that must be asked by water utilities—what are you doing with water saved? If you are selling it for future growth rather than developing new supplies or building a new reservoir, it means one thing. If you are saving it in a reservoir in case the current drought lasts longer, it is another thing entirely.

Some conservation measures are designed to be temporary in nature simply to allow the utility and its customers to get through a current drought. The behavior called for is not expected to continue longer than the drought. Other conservation measures are designed to permanently alter behavior or change the hardware involved with water use. Permanent conservation measures give the utility a choice of whether to sell its conserved water for future growth, save it in a reservoir, or leave it in a stream.

Drought-response measures may in fact be viewed by customers as something of a sacrifice, and they are certainly viewed by the utility as temporary. A utility would be foolish to sell water conserved as a drought measure for future growth. However, if the utility sells all the water conserved by customers pursuant to a permanent water conservation program, they are subject to demand hardening. This means that conservation savings in a future drought will be difficult. If the utility serves two thousand people with the water previously used to serve one thousand, it looks like a great victory. When the next drought hits, the utility will ask customers to cut back by 30 or 40 percent more. They may not be able to do so, or they will find doing so extremely difficult, and the utility would have succeeded only in hardening the demand. The flexibility previously available to the utility to deal with a drought is gone.

At the end of the day, all utilities must analyze conservation measures to understand their cost, their yield in acre-feet saved per dollar expended, and their effect on the customer, in the operational capability and flexibility of the utility, and on downstream owners of water rights generated by return flows.

There are many who believe that the answer to any water supply dilemma in the United States is more conservation. However, it is not that simple.

6. Water lawyers and the environmental community, although always professing to help, make the job of a water utility manager infinitely more complicated and frustrating.

The environmental community, of course, is not of one mind on any subject, and so characterizing them all in a sentence or two is not legitimate. But I do recall a meeting a few months ago with a statewide community-based self-selected group called the Standing Committee on Environmental and Water Interests. This committee is a special interest group that convenes to pursue a particular environmental interest if there is no preexisting special interest group empowered to pursue that interest. It monitors public utterances on water use, pharmaceuticals in drinking water, forest fires and watersheds, water conservation, low-flush toilets, and development plans in order to see who might be offended, and then it takes offense if no one else has the time or inclination. It watches water and utility power structures and petitions for redress, regulatory enforcement, minimum stream flow, basin of origin protection, and other assorted good causes. It is an extraordinarily hard-working group, never at rest, always vigilant. Recently—and listen to this list—the committee has taken up the cause of the inequality of water distribution among the western states, the preservation of sagebrush in western Colorado, the conversion of all Forest Service land to national parks, the abolition of smoking in football stadiums, the establishment of causal connection between lawn irrigation and global warming, the proliferation of phreatophytes, and the application of integrated-resources planning techniques to global overpopulation. They wanted to know what I was going to do to further their cause.

The water lawyers are of course a different group entirely. Some water lawyers also believe they are environmentalists, although they must not have examined closely how many trees have been cut down to further their complex conditional clauses, subparagraphs, and footnotes. The Colorado River Compact was written in 1922 and it is one of the most important documents affecting any water user taking water from the Colorado River or its tributaries. The compact is about ten pages long. I hesitate to think how long it would be or how many years it would have taken to negotiate had the current stable of Colorado water lawyers been involved in its negotiation. The ten-page document would be at least six hundred pages in length, but it would not be more clear than the ten-page version. This imagined effort by the water lawyers to solve every conceivable condition or consequence of water flow or water use along the Colorado River for the next hundred years would spawn a dozen or more subsidiary questions, each leading to further paragraphs and subparagraphs of "clarifying" language. At the end of the day we should be able to rely on common sense, good faith, and mutually agreed upon intentions. In the water world that I occupy, that seldom now seems to be the case, and simple documents take years to negotiate and forests of trees for the paper to print them on.

7. The water world has been taken over by the computer geeks, here as much as elsewhere in the world.

It is a good thing, as manager of a water utility, that I don't have to know a lot about the interaction between computer models, hydrology, water-treatment plants, and customer bills. As far as computers are concerned, I thought a window was something that you had to clean and that memory was something you lost with age. I've always believed that if you had a three-and-a-half-inch floppy, you hoped that nobody found out about it, and that backup was something you worried about in your wastewater system. As far as I know, a cursor uses profanity and a keyboard is part of a musical instrument.

8. Most of the world does not understand that water is not a finite resource, and that we are not "using it up."

Water is not like gasoline, or coal, or—God forbid—oil. Water is a renewable resource. The hydrologic cycle is what makes it so. Turning off the water while you brush your teeth does not save water from disappearing forever into outer space but simply means the utility will not have to capture and treat a couple of extra gallons of water for you that day. If you do "waste" those few gallons, an entity downstream will get them eventually, and their use or a subsequent use will put the water into the hydrologic cycle, where it will eventually return as rain or snow somewhere. In the end, water conservation is not about waste, but about efficiency, and about understanding that the benefits of water conservation are all local and not global.

Most of the public is ignorant of some of the basics about water, water supply, the hydrologic cycle, the cost and techniques of water treatment, etc. In this part of the world, obtaining the water to put in your faucet is an engineering, legal, and political feat of some complication. When I explain to people that they can save some water by flushing their toilet with a bucket, many ask, "How do I get the water into the tank?" I have found that 40 percent of the public in Colorado does not know that pouring a bucket of water into the toilet bowl will make it flush. Let's start our water education with that thought, and finish this talk with it as well.

—Chips Barry
Manager, Denver Water, 1991–2010

Notes

Introduction

1. Martin V. Melosi, *The Sanitary City: Urban Infrastructure in America from Colonial Times to the Present* (Baltimore: Johns Hopkins Univ. Press, 2000), 1.
2. Daniel Tyler, *The Last Water Hole in the West: The Colorado–Big Thompson Project and the Northern Colorado Water Conservancy District* (Niwot: Univ. Press of Colorado, 1992), 5.
3. Diane Raines Ward, *Water Wars: Drought, Folly, and the Politics of Thirst* (New York: Riverhead/Penguin Putnam, 2002), 72.
4. Earl Pomeroy, *The American Far West in the Twentieth Century*, edited by Richard Etulain (New Haven, CT: Yale Univ. Press, 2008), 569.
5. Quotation from Steven Solomon, *Water: The Epic Struggle for Wealth, Power, and Civilization* (New York: Harper Collins, 2010), 297; Gerald T. Koeppel, *Water for Gotham* (Princeton, NJ: Princeton Univ. Press, 2001); Sarah Elkind, *Bay Cities and Water Politics: The Battle for Resources in Boston and Oakland* (Lawrence: Univ. Press of Kansas, 1998); Michael Rawson, *Eden on the Charles: The Making of Boston* (Cambridge, MA: Harvard Univ. Press, 2010).
6. Martin Reuss and Stephen H. Cutcliffe, eds., *The Illusory Boundary: Environment and Technology in History* (Charlottesville: Univ. of Virginia Press, 2010), 9, 266, 291.
7. Reuss and Cutcliffe, *Illusory Boundary*, 21–22, 72.
8. Edward Abbey, *Desert Solitaire: A Season in the Wilderness* (1968); reprinted (New York: Touchstone/Simon and Schuster, 1990), 126.
9. "Will Our Rivers Survive?" a Roundtable Forum with Bruce Hallin, Patricia Mulroy, Peter Gleick, Chips Barry, Amy Souers Kober, and Peter Grubb, *Sunset Magazine* (Feb. 2, 2008).

Part One

Chapter One

Engineered Eden

1. Thomas Hornsby Ferril, *Thomas Hornsby Ferril and the American West*, eds., Robert C. Baron, Stephen J. Leonard, and Thomas J. Noel (Golden, CO: Fulcrum Publishing and the Center of the American West, 1996), 36.
2. Richard Hugo, *Making Certain it Goes On: The Collected Poems of Richard Hugo*, ed. William Kittredge (New York: WW Norton, 1984).
3. Kirk R. Johnson and Robert G. Raynolds, *Ancient Denvers: Scenes from the Past 300 Million Years of the Colorado Front Range* (Denver: Denver Museum of Nature & Science, 2002), 1–2.
4. Elliott West, *The Contested Plains: Indians, Goldseekers, and the Rush to Colorado* (Lawrence: Univ. Press of Kansas, 1998), 40; John H. Moore, *The Cheyenne Nation: A Social and Demographic History* (Lincoln: Univ. of Nebraska Press, 1987), 153–54.
5. Zebulon Montgomery Pike, *The Expeditions of Zebulon Montgomery Pike*, Vol. II, ed. Elliott Coues (New York: F. P. Harper, 1895), 525.
6. William Goetzmann, *Exploration and Empire: The Explorer and the Scientist in the Winning of the American West* (1966); reprinted (New York: Vintage Books, 1972), 51, 62.
7. Edwin James, *Account of an Exploring Expedition from Pittsburgh to the Rocky Mountains*, Vol. II (London: Longman, Hurst, Rees, Orme, and Brown, 1823); reprinted (Readex Microprint, 1966), 311.
8. Francis Parkman, *The Oregon Trail*, 4th ed. (Boston: Little, Brown, and Co., 1872), 291.

9. Parkman, *Oregon Trail,* 291–93, 296.
10. Leroy R. Hafen, *Colorado: A Story of the State and Its People* (Denver: The Old West Publishing Company, 1948), 141.
11. Elliott West, *The Contested Plains,* 106–7; Hafen, *Colorado,* 142–43.
12. Gunther Barth, *Instant Cities: Urbanization and the Rise of San Francisco and Denver* (New York: Oxford Univ. Press, 1975).
13. Robert Perkin, *The First Hundred Years: An Informal History of Denver and the Rocky Mountain News* (Garden City, NY: Doubleday & Company, 1959), 210–21.
14. The Goldrick quotation—indeed, the entire length of Professor Goldrick's lyric article on the calamity—appears in Perkin, *The First Hundred Years,* 218–19; Clyde King, *The History of the Government of Denver with Special Reference to its Relations with Public Service Corporations* (Denver: Fisher Book Company, 1911), 50–51.
15. Albert Richardson, *Beyond the Mississippi: From the Great River to the Great Ocean; Life and Adventures on the Prairies, Mountains, and Pacific Coast* (Hartford, CT: American Publishing Company, 1869), 177, 179.
16. Quoted in Stephen Leonard and Thomas Noel, *Denver: Mining Camp to Metropolis* (Niwot: Univ. Press of Colorado, 1990), 27.
17. William Gilpin, *Mission of the North American People, Geographical, Social, and Political;* revised 2nd ed. (Philadelphia: J. B. Lippincott, 1873), 71, 119, 72, 76. Originally published in 1860 as *The Central Gold Region.*
18. Gilpin, *Mission,* 120, 49.
19. Charles Dana Wilber, *The Great Valleys and Prairies of Nebraska and the Northwest* (Omaha: Daily Republican Printing, 1881), 68, 69, 71.
20. King, *History of the Government of Denver,* 65.
21. Earl L. Mosley, *History of the Denver Water System, 1858–1919,* Vol. I (Denver: Board of Water Commissioners, 1962), 25–27.
22. Mosley, *History of the Denver Water System,* 26–29.
23. Mosley, *History of the Denver Water System,* 33–35, 38.
24. Mosley, *History of the Denver Water System,* 55–64; Louisa Ward Arps, *Denver in Slices: A Historical Guide to the City* (Denver: Sage Books, 1959), 65–72.
25. Mosley, *History of the Denver Water System,* 118.
26. Mosley, *History of the Denver Water System,* 131–32; Leonard and Noel, *Denver,* 49; King, *History of the Government of Denver,* 39; Arps, *Denver in Slices,* 65–72; "Denver Water—A Condensed History" in 2005 Budget Book, Denver Water, http://denver water.org (accessed 5/28/06).
27. *Rocky Mountain News* and Henry Leach quoted in Mosley, *History of the Denver Water System,* 75; Rezin H. Constant, quoted in Lyle Dorsett, *The Queen City: A History of Denver* (Boulder, CO: Pruett Publishing Company, 1977), 90; Crawford quoted in Eugene Berwanger, *The Rise of the Centennial State: Colorado Territory, 1861–1876* (Urbana: Univ. of Illinois Press, 2007), 14.
28. Richardson, *Beyond the Mississippi,* 333; "A Trip to Colorado, Part VI.—Farming in Colorado," *New York Tribune* article reprinted in the *Rocky Mountain News,* August 29, 1866, 2; William Makepeace Thayer, *Marvels of the New West: A Vivid Portrayal of the Stupendous Marvels in the Vast Wonderland West of the Missouri River* (Norwich, CT: Henry Bill Publishing Company, 1888), 355–56.
29. "How a Man on the Water Wagon Solved a Nineteenth Century Mystery," *Denver Water News* 11, no. 8 (August 1943): 3; Arps, *Denver in Slices,* 68.
30. "Water Works: What the Next Improvement Should Be," *Rocky Mountain News,* November 22, 1870; James L. Cox, *The Development and Administration of Water Supply Programs in the Denver Metropolitan Area* (PhD dissertation, Univ. of Colorado–Boulder, Department of Political Science, 1965), 185.
31. "Water Works: What the Next Improvement Should Be," *Rocky Mountain News,* November 22, 1870; "We Want More Water," *Rocky Mountain News,* June 9, 1871.
32. "We Want More Water," *Rocky Mountain News,* June 9, 1871.

33. Dorsett, *Queen City*, 77–79; King, *History of the Government of Denver*, 79–82.
34. King, *History of the Government of Denver*, 82; Mosley, *History of the Denver Water System*, 213.
35. Gretchen Claman, "A Typhoid Fever Epidemic and the Power of the Press in Denver in 1879," *The Colorado Magazine* 56 (Summer–Fall 1979): 143–60.
36. "Analysis of the Water: An Experienced Chemist's Diagnosis of the Latest Night–Mare—We Have the Purest and Best Drinking Water in the World," *Rocky Mountain News*, July 15, 1879.
37. John Willmuth Hill, *The Purification of Public Water Supplies* (New York: The D. Van Nostrand Co., 1898), 104–6; *Rocky Mountain News*, December 12, 1879; Claman, "A Typhoid Fever Epidemic," 143–60.
38. Claman, "A Typhoid Fever Epidemic," 159; Dieter H. M. Groschel and Richard B. Hornick, "Who Introduced Typhoid Vaccination: Almroth Wright or Richard Pfeiffer?" *Reviews of Infectious Diseases* 3, no. 6 (November–December 1981): 1251–54.
39. "An Open Letter to Colonel Archer from Mr. Davis," and "Colonel Archer's Reply to the M.D.s," *Rocky Mountain News*, December 13, 1879.
40. Arps, *Denver in Slices*, 49; James D. Schuyler, "The Water Supply and Water Works of Denver Colorado," Denver, CO: February 1891, 1–2, in James D. Schuyler Papers, Folder 152.4, Water Resources Center Archives, Univ. of California–Berkeley; King, *History of the Government of Denver*, 141–45; Jerome C. Smiley, *History of Denver: With Outlines of the Earlier History of the Rocky Mountain Country* (Denver: The Times-Sun Publishing Company, 1901), 794–96.
41. Smiley, *History of Denver*, 795; Schuyler, "The Water Supply and Water Works of Denver Colorado," 1–2; Mosley, *History of the Denver Water System*, 274.
42. Schuyler, "The Water Supply and Water Works of Denver Colorado,"1–3; Smiley, *History of Denver*, 797–98.
43. Carl Abbott, Stephen Leonard, and Thomas Noel, *Colorado: A History of the Centennial State* (Niwot: Univ. Press of Colorado, 2005), 91–93; Mosley, *History of the Denver Water System*, 114, 149.
44. *Rocky Mountain News* editorial comment, August 27, 1871, quoted in Mosley, *History of the Denver Water System*, 132–37.
45. Mosley, *History of the Denver Water System*, 118–29.
46. Mosley, *History of the Denver Water System*, 136–42; King, *History of the Government of Denver*, 83.
47. King, *History of the Government of Denver*, 82–85; Arps, *Denver in Slices*, 47–49.
48. Dorsett, *Queen City*, 79; Walter Eha, *Water for Denver: How Water Helped Build a City* (Denver: Denver Water/Denver Public Library, 2006), 9. The original documents were printed in 1944, and can be found in the Walter Eha Papers, WH115, Western History Collection of the Denver Public Library.
49. Eha, *Water for Denver*, 9; King, *History of the Government of Denver*, 148; Schuyler, "The Water Supply and Water Works of Denver Colorado," 5; Smiley, *History of Denver*, 799.
50. James D. Schuyler, "Report on the Water Supply for 1892 and 1893, and the Way to Secure It," Denver, CO, February 16, 1892, 1–2, in James D. Schuyler Papers, Folder 152.5, Water Resources Center Archives, Univ. of California–Berkeley; King, *History of the Government of Denver*, 147; Smiley, *History of Denver*, 799; Eha, *Water for Denver*, 10.
51. Mosley, *History of the Denver Water System*, 512; Leonard Metcalf and George G. Anderson, "Inventory of the Property of The Denver Union Water Co., Denver, Colorado, as of October 31, 1913" (Denver: Smith-Brooks Press, 1914), 9.
52. King, *History of the Government of Denver*, 149, 152.
53. King, *History of the Government of Denver*, 152, 201.
54. Steven Frederick Mehls, *David H. Moffat Jr.: Early Colorado Business Leader* (PhD dissertation, Univ. of Colorado–Boulder, Department of History, 1982), iv.
55. Edgar C. McMechan, *Walter Scott Cheesman: A Pioneer Builder of Colorado* (193–?), 10.

56. Jane G. Haigh, "Interests and the Political Economy of Corruption," in *Political Power, Patronage, and Protection Rackets: Con Men and Political Corruption in Denver, 1889–1894* (PhD dissertation, Univ. of Arizona, 2009), 11–12.
57. Richard White, "Information, Markets, and Corruption: Transcontinental Railroads in the Gilded Age," *Journal of American History* 90, no. 1 (June 2003): 19.
58. White, "Information, Markets, and Corruption," 20; Haigh, *Political Power, Patronage, and Protection Rackets*, 11.
59. David Wrobel, *Promised Lands: Promotion, Memory, and the Creation of the American West* (Lawrence: Univ. of Kansas Press, 2002), 2.
60. McMechan, *Cheesman*, 8.
61. Haigh, *Political Power, Patronage, and Protection*, 2; Mehls, *David H. Moffat Jr.*, 156, iv.; Thomas E. Cronin, "Coloradans Are a Skeptical Bunch," *The Denver Post*, May 16, 2010.
62. Smiley, *History of Denver*, 808.
63. Smiley, *History of Denver*, 808.

Chapter Two

Go Take It from the Mountain

1. Martin V. Melosi, *The Sanitary City: Urban Infrastructure in America from Colonial Times to the Present* (Baltimore: Johns Hopkins Univ. Press, 2000), 148.
2. Carl Abbott, *How Cities Won the West: Four Centuries of Urban Change in Western North America* (Albuquerque: Univ. of New Mexico Press, 2008), 152, 156.
3. For illuminating histories of the sagas to secure municipal water in Los Angeles and San Francisco, see William L. Kahrl, *Water and Power: The Conflict over Los Angeles' Water Supply in the Owens Valley* (Berkeley: Univ. of California Press, 1982), and Robert W. Righter, *The Battle over Hetch Hetchy: America's Most Controversial Dam and the Birth of Modern Environmentalism* (New York: Oxford Univ. Press, 2005).
4. Earl L. Mosley, *History of the Denver Water System, 1858–1919*, Vol. II (Denver: Board of Water Commissioners, 1962), 596–99; Hiram M. Chittenden, "Examination of Reservoir Sites in Wyoming and Colorado," in *Annual Reports of the War Department, Fiscal Year Ended June 30, 1898, Report of the Chief of Engineers*, Part 4, (Washington: US Government Printing Office, 1898), 2837.
5. Clyde King, *The History of the Government of Denver with Special Reference to its Relations with Public Service Corporations* (Denver: Fisher Book Company, 1911), 82–83.
6. Martin Reuss, "Probability Analysis and the Search for Hydrologic Order in the United States, 1885–1945," *Water Resources Impact* 4, no. 3 (May 2002): 7–15; Martin Reuss, personal communication with the authors, June 8, 2010; Carroll Pursell, *The Machine in America: A Social History of Technology*, 2nd ed. (Baltimore: Johns Hopkins Univ. Press, 2007), 131; Edwin T. Layton Jr., *The Revolt of the Engineers: Social Responsibility and the American Engineering Profession* (Cleveland: The Press of Case Western Reserve Univ., 1971), viii; Martin Reuss, "The Corps of Engineers and Water Resources in the Progressive Era (1890–1920)," *Public Works History* 28 (Public Works Historical Society, 2009), 1.
7. "Denver Water Is Defense Against Typhoid," *Denver Water News*, September 1941; "History of Denver Water," www.denverwater.org/About Us/History (accessed October 28, 2009); Jerome C. Smiley, *History of Denver: With Outlines of the Earlier History of the Rocky Mountain Country* (Denver: The Times-Sun Publishing Company, 1901), 802; "The Water System of Denver," *Frank Leslie's Illustrated Weekly*, January 2, 1892.
8. James D. Schuyler, "Report of Consulting Engineer to the President and Director of the Citizens' Water Company," Denver, CO, April 25, 1890, p. 2; James D. Schuyler Papers, Water Resources Center Archives, Schuyler Folder 152.3, Univ. of California–Berkeley.

9. Walter Eha, *Water for Denver: How Water Helped Build a City* (Denver: Denver Water/ Denver Public Library, 2006), 12–13. Eha reported that the city used 8,536,165,582 gallons of water in 1896. The 33,859 population figure from the US Census: 1900. Also see "Lake Cheesman: A Chronological Memorandum." Both are located in Box 393-203576, Denver Water Records and Document Administration.
10. Engineering Vault Photos, Book One and Book Two: Cheesman Dam, Denver Water Records and Document Administration. Photo No. 65 from Book One, p. 58, is a blueprint with precise measurements of the proposed dam. Also see "South Platte Reservoir Site," *Denver Republican*, April 4, 1898; "Highest Dam in the World to Insure Denver Against Water Famine," *Denver Republican*, July 2, 1899; "Letter from Colorado State Engineer Addison J. McCune to South Platte Canal and Reservoir Company Chief Engineer C. P. Allen," March 13, 1900, located in Box 393-203576, Denver Water Records and Document Administration; and memo from George Andersen, "Consultation in Regard to Construction of Goose Creek Reservoir," January 20, 1899, located in Box 393-203576, Denver Water Records and Document Administration.
11. Engineering Vault Photos, Book One and Book Two: Cheesman Dam, Denver Water Records and Document Administration; "Highest Dam in the World to Insure Denver Against Water Famine," *Denver Republican*, July 2, 1899.
12. Eha, *Water for Denver*, 12–13, and Engineering Vault Photos, Book Two: Cheesman Dam (especially photos No. 88, p. 67; No. 206, p. 74; and No. 95, p. 71).
13. William Flick quoted in "Swept by Million Tons of Water," *Rocky Mountain News*, May 4, 1900; The story about Zebulon Swan comes from Glenn Jenkins, Denver Water History Project, interviewed by Jane Earle, DVD, courtesy of Denver Water Board.
14. "Swept by Million Tons of Water," *Rocky Mountain News*, May 4, 1900.
15. "Swept by Million Tons of Water," *Rocky Mountain News*, May 4, 1900.
16. Charles Harrison quoted in Clifford Allen Betts, Clifford Johnson, and John S. Marshall, "Nomination of Cheesman Dam for Designation as a National Historic Civil Engineering Landmark," report submitted through the Colorado Section of the American Society of Civil Engineers (April 1972), 4, ii, 6.
17. Charles Bjork, "Experiences in the Building of Cheesman Dam," *The Colorado Magazine* 26, no. 2 (April 1949): 98–104.
18. Betts, Johnson, and Marshall, "Nomination," 11.
19. Stephen Leonard and Thomas Noel, *Denver: Mining Camp to Metropolis* (Niwot: Univ. Press of Colorado, 1990), 140.
20. Betts, Johnson, and Marshall, "Nomination," ii–iii.
21. Jeffrey Kastner, editor, *Land and Environmental Art* (London: Phaidon Press, 1998), Preface, 11, 15.
22. Kastner, *Land and Environmental Art*, 11.
23. Smiley, *History of Denver*, 807–8.
24. Richard White, personal communication to author, July 11, 2010.
25. Melosi, *The Sanitary City*, 123.
26. Information about the original franchise—between the city and the consortium of companies that became Denver Union—can be found in Clyde King, *The History of the Government of Denver*, 148–52. The franchise contract obligated the water company to "fix schedule rates for private consumers equivalent to the average rate prevailing in the cities of Chicago, St. Louis, and Cincinnati," to provide water that was "of a quality as good and fit for private consumption," and after twenty years, to allow the city to purchase the company if it saw fit.
27. King, *History of the Government of Denver*, 204–6. For a concise recapitulation of negotiations between the water company and the city from the stockholders' perspective, see W. N. W. Blayney, E. S. Kassler, W. F. Hayden, J. H. Porter, the Stockholders Protective Committee, "To the Shareholders of the Denver Union Water Company,"

September 9, 1918, in the Record of Minutes for the Stockholders Protective Committee, Denver Water Records and Document Administration.

28. "Proposition from Water Company: It Offers to Sell its Water Plant to the City of Denver," *Denver Republican*, May 18, 1898.
29. "Must Have All or Nothing: Water Company Furnishes a List of Property but Declines to Attach Detailed Valuation of Items," *Rocky Mountain News*, September 1, 1899.
30. "Judge Riner Decides Water Case against the City of Denver," *Rocky Mountain News*, February 5, 1901. Specific correspondence between the city and the water company between 1899 and 1900, concerning the sale of its assets, can be found in File 264 of the Secretary Files, Denver Water Records and Document Administration. The city engineer appraised the value of Denver Union's system at $3,763,617, to be exact. See also King, *History of the Government of Denver*, 205–6.
31. Armour C. Anderson for the Public Utilities Commission, "Report of the Public Utilities Commission of the City and County of Denver for the Year 1910," (Denver: Public Utilities Commission, December 31, 1910), 5–8.
32. King, *History of the Government of Denver*, 285; "Letter from David Moffat to Mssrs. Underwood, Van Vorst, and Hoyt," August 13, 1910, Box 371-693058, Denver Water Records and Document Administration.
33. King, *History of the Government of Denver*, 281; Leonard and Noel, *Denver*, 128–39. Robert Speer spearheaded the business-friendly City Charter campaign to the chagrin of Progressives, who favored a charter with provisions for municipal ownership. Speer was elected mayor shortly after the charter was adopted. Progressives often referred to the collusion between public officials and corporations as "the beast." Included in this group were corporations with a large stake in obtaining city franchises, like the Denver City Tramway Company, the Gas and Electric Company, and the Denver Union Water Company.
34. Anderson, "Report of the Public Utilities Commission of the City and County of Denver for the Year 1910," 5–8.
35. King, *History of the Government of Denver*, 285–87; Anderson, "Report of the Public Utilities Commission," 8; "Municipal Ownership of Water-Works at Denver, Colorado," *Engineering News* 63, No. 22 (June 2, 1910): 659; Charter of the City and County of Denver (Denver: Smith-Brooks Printing Company, 1911), 93–97.
36. Internal documents indicate that the bondholders of Denver Union (particularly the New York Trust Company) and company directors differed in how they proposed to respond to the city's franchise rejection; it appears that the bondholders, not the company itself, instigated the litigation that led to the US Supreme Court case *City and County of Denver v. New York Trust Company*. The case was argued on October 28–29, 1912, and was decided on May 26, 1913. For more information, see "Letter from Underwood, Van Vorst, and Hoyt to Denver Union Water Company," August 13, 1910; "Letter from David Moffat to Mssrs. Underwood, Van Vorst, and Hoyt," August 15, 1910; and "Western Union Telegram from Underwood, Van Vorst, and Hoyt to Denver Union Water Company," August 18, 1910, in Box 371-693058, Denver Water Records and Document Administration.
37. Eha, *Water for Denver*, 18; Denver Union Water Company Board Meeting Minute Record Book, Volume 2, December 8, 1910, Denver Water Records and Document Administration; *City and County of Denver v. New York Trust Company*, 229 US 123, Nos. 642 and 643 (1913).
38. A. Lincoln Fellows, C. P. Allen, E. C. Van Diest, "Report to the Public Utilities Commission of the City and County of Denver" (Denver Water Records and Document Administration: January 16, 1914, 2; Mosley, *History of the Denver Water System*, 672.
39. Eha, *Water for Denver*, 19–20; Fellows, Allen, and Van Diest, "Report to the Public Utilities Commission of the City and County of Denver"; Mosley, *History of the Denver Water System*, 701. To be exact, the Board of Engineers report appraised Denver

Union's total value at $10,044,778 in 1914. In 1916, the Van Sant-Houghton firm estimated that it would cost the city at least $27,479,498 to build its own system.

40. *Denver Union Water Company v. City and County of Denver*, No. 6274, District Court of the United States, District of Colorado, report submitted by W. J. Chinn on October 5, 1915, and endorsed by District Judge Robert E. Lewis on January 28, 1916; W. N. W. Blayney et al., "To the Shareholders of the Denver Union Water Company."
41. "Minutes of Meeting of the Public Utilities Commission of the City and County of Denver," February 21, 1916, Denver Water Records and Document Administration; Eha, *Water for Denver*, 20; Mosley, *History of the Denver Water System*, 672; *City and County of Denver v. Denver Union Water Company*, 246 US 178, Nos. 294 and 295 (1918).
42. "Minutes of Meeting of the Public Utilities Commission," February 21, 1916; Eha, *Water for Denver*, 20; *City and County of Denver v. Denver Union Water Company*, 246 US 178, Nos. 294 and 295 (1918).
43. On structure and powers of DWB, see James L. Cox, *The Development and Administration of Water Supply Programs in the Denver Metropolitan Area* (PhD dissertation, Univ. of Colorado–Boulder, Department of Political Science, 1965), 190–212; and "About Denver Water," at www.denverwater.org.
44. City and County of Denver, "Charter of the City and County of Denver" (1960 Compilation), Article 4, Section 26. For an overview of the responsibilities of Denver Water, see Denver Water, "A Refreshing Look at Denver Water" (ca. 2007).
45. Eha, *Water for Denver*, 16; and "Capitol Hill Reservoir History" at www.denverwater.org.
46. Eha, *Water for Denver*, 13; and "Water Quality" at www.denverwater.org.
47. US Bureau of the Census, 1920, cited in Richard L. Forstall, ed., "Colorado: Population of Counties by Decennial Census, 1900 to 1990" (Washington, DC: US Bureau of the Census, 1995), available online at www.census.gov/population/cencounts/co190090.txt.
48. *Rocky Mountain News*, June 8, 1921; *Rocky Mountain News*, June 9, 1921; US Geological Survey, Department of the Interior, Floods in Colorado (Washington, DC: US Government Printing Office, 1948), 27; State of Colorado, Twenty-First Biennial Report of the State Engineer to the Governor of Colorado for the Years 1921–1922 (Denver: Eames Brothers, 1923), 33–34.
49. Eha, *Water for Denver*, 21.
50. Abbott, *How Cities Won the West*, 155.

Part Two

Chapter Three

Water Development: "The Plot Thickens"

1. Bert Hanna, "Colorado Water War: Denver's Good Faith Questioned by Western Slope Interests," *The Denver Post*, January 22, 1964.
2. Saunders quoted in Lee Olson, "The Arid West: Water Use Will Determine How, Where People May Live," *The Denver Post*, May 21, 1967.
3. Glenn G. Saunders, "Reflections on Sixty Years of Water Law Practice," Natural Resources Law Center Occasional Papers Series (Boulder: Univ. of Colorado School of Law, 1989), 1.
4. Mark Obmascik, "Dean of Colorado Water Lawyers Dies," *The Denver Post*, May 3, 1990.
5. Saunders quoted in Tom Gavin, "Denver Keeps Alive Hope for Blue River," *Rocky Mountain News*, December 14, 1954.
6. William Nelson, "Talk Secret Water Programs," *Grand Junction Sentinel*, July 3, 1955; "Usual Ruthless Tactics," *Grand Junction Sentinel*, September 8, 1963.
7. Bert Hanna, "Legal Threats Posed to Denver Water Diversion Plan," *The Denver Post*, August 11, 1963; James O. Wood and Gene Cooper, "City Defies US in Water

Showdown," *Rocky Mountain News*, September 2, 1963; Bert Hanna, "Trial Delay Raises Hopes of Out-of-Court Dillon Settlement," *The Denver Post*, January 19, 1964.

8. Quoted in Bill Sonn, "Water Board Lawyer Attacks Indians as Inferior," *Straight Creek Journal*, June 4–11, 1974.
9. "Blue River Project to Triple Water Supply," *Rocky Mountain News*, November 26, 1944; Dick Prouty, "West Slope Reservoir Concept Studied," *The Denver Post*, November 28, 1971.
10. John Opie, *Nature's Nation: An Environmental History of the United States* (Fort Worth, TX: Harcourt Brace College Publishers, 1998), 323.
11. Marc Reisner, *Cadillac Desert: The American West and Its Disappearing Water* (New York: Viking, 1986), 450–51.
12. Morgan Lawhon, "Population Pressure Called Threat to US Beauty," *The Denver Post*, September 26, 1965.
13. "Leave Granted to Saunders by Water Board," *The Denver Post*, September 29, 1965.
14. Olson, "In the Arid West."
15. Bert Hanna, "Water Counsel Eased Aside," *The Denver Post*, February 23, 1966; Obmascik, "Dean of Colorado Water Lawyers Dies."
16. Editorial, *Grand Junction Sentinel*, March 2, 1966.
17. "Retirement of Glenn G. Saunders" and "Supplemental Information," prepared for the consideration of the Denver Water Board by Manager R. B. McRae, August 26, 1969. Located in the personal scrapbooks of Glenn G. Saunders.
18. "Water Attorney Will Retire," *The Denver Post*, August 27, 1969; Jeff Rosen, "Transmountain Diversions Challenged in Court Suit," *Rocky Mountain News*, December 1, 1973; Suzanne Weiss, "Water Suit Testimony Heard," *Rocky Mountain News*, September 10, 1975.
19. "Usual Ruthless Tactics," editorial, *Grand Junction Sentinel*, September 8, 1963. Also see Stephen J. Leonard and Thomas J. Noel, *Denver: Mining Camp to Metropolis* (Niwot: Univ. Press of Colorado, 1990). In another history of Denver, Lyle W. Dorsett makes only the briefest mention of Saunders, referring only to his work in the 1930s: "City attorney Malcolm Lindsey and his astute assistant Glenn Saunders handled the legal problems to securing water rights" during the administration of Denver Mayor Ben Stapleton. Lyle W. Dorsett, *The Queen City: A History of Denver* (Boulder, CO: Pruett Publishing, 1977), 207.
20. "Retirement of Glenn G. Saunders" and "Supplemental Information," prepared for the consideration of the Denver Water Board by Manager R. B. McRae, August 26, 1969; Board of Water Commissioners, "Resolution" regarding the retirement of Glenn G. Saunders, September 15, 1969. Located in the personal scrapbooks of Glenn G. Saunders.
21. "Water Board Counsel Hits Wilderness Plan," *Rocky Mountain News*, November 22, 1958; John Morehead, "Rationing 'Almost Sure,' Water Chief Claims," *The Denver Post*, May 1, 1975.
22. Bert Hanna, "Dillon Bid Pleases Saunders," *The Denver Post*, December 4, 1963.
23. Saunders, "Reflections on Sixty Years," 4.
24. Jack Ross, "Six of the Greatest: Glenn G. Saunders," *The Colorado Lawyer* 34, no. 7 (July 2005): 43.

Chapter Four

Dealing in Diversions

1. Kenneth Chang, "Scientists See Fresh Evidence of More Water on the Moon," *The New York Times*, March 9, 2010; "Water on the Moon," *Science News*, January 2, 2010.
2. Thomas Farnham, *Travels in the Great Western Prairies, the Anahuac and Rocky Mountains, and in the Oregon Territory*, Vol. I (London: Richard Bentley, 1843) 258, 261; John Charles Frémont, *The Expeditions of John Charles Frémont*, Vol. I, ed. Donald

Jackson and Mary Lee Spence (Urbana: Univ. of Illinois Press, 1970), 715–16; Bayard Taylor, *Colorado: A Summer Trip* (New York: G. P. Putnam and Son, 1867), 88, 100.

3. Taylor, *Colorado,* 90, 92.
4. Taylor, *Colorado,*105, 111.
5. University of Virginia, "Historical Census Browser," *University of Virginia, Geospatial and Statistical Data Center,* available online at http://fisher.lib.virginia.edu/collections/stats/histcensus/index.html; Colorado Water Conservation Board, *Statewide Water Supply Initiative Report* (Denver: Colorado Water Conservation Board, 2004).
6. "Denver Water Claims Hit at Hearing," *The Denver Post,* November 20, 1959; David Barnard Cole, "Transmountain Water Diversion in Colorado," *The Colorado Magazine* 23, no. 2 (March 1948): 53; "Irrigation of Agricultural Lands," original data from the *Sixteenth Census of the United States, 1940,*184.
7. Norris Hundley Jr., "The West Against Itself: The Colorado River—An Institutional History," in *New Courses for the Colorado,* ed. Gary Weatherford and F. Lee Brown (Univ. of New Mexico Press, 1996). The text of the Colorado River Compact is available online through the US Bureau of Reclamation, www.usbr.gov.
8. Earl L. Mosley, *History of the Denver Water System, 1858–1919,* Vol. 6 (Denver: Board of Water Commissioners, 1962), 1, 122; Walter Eha, *Water for Denver: How Water Helped Build a City* (Denver: Denver Water/Denver Public Library, 2006), 21–22.
9. "In the Twenties, Denver Turned to the Western Slope for Water," *Denver Water News* 10, no. 7 (July 1942); Denver Board of Water Commissioners Meeting Minutes from June 10, 1925; Eha, *Water for Denver,* 24.
10. "Denver Water Supply Falling," *Rocky Mountain News,* November 11, 1932.
11. Thomas B. McKee, et al., *A History of Drought in Colorado: Lessons Learned and What Lies Ahead* (Fort Collins: Colorado Water Resources Research Institute, 2000), 15; "School Closed Because of Dust: Eastern Colorado Suffers Severely," *Rocky Mountain News,* March 9, 1935; "Dust Hurricane Spreads Damage in Four States," *Rocky Mountain News,* March 16, 1935; "Three Perish as Duststorm Strikes Denver," *Rocky Mountain News,* March 27, 1935.
12. Maurice Leckenby, "State Water Famine Worst in Many Years," *Rocky Mountain News,* September 25, 1932; "Denver Water Supply Falling," *Rocky Mountain News,* November 6, 1932; Denver Water, *Your Denver Municipal Water System: Where the City's Water Comes From—How it Is Purified—How It Gets into Your Home* (Denver: Denver Water, 1934), 23.
13. "Public is Cooperating in Saving Water," *Denver Water News* 3, no. 1 (January 1, 1935): 2; "Denver Reports Record Low Water Consumption," *The Denver Post,* December 9, 1935; "Water Usage at New Low," *Rocky Mountain News,* February 5, 1938; "Sprinkling Permitted on Two Days Each Week, During Daylight Hours," *Denver Water News* 3, no. 5 (May 1, 1935): 1; "Board Thanks Customers for Conservative Use of Water," *Denver Water News* 6, no. 11 (November 1938): 1; "City of Beautiful Lawns" quoted in *Rocky Mountain News,* September 25, 1932; the claim that "Denver has more and greener lawns than any other city of its size in the world" was made in *Open Door,* a publication of the Industrial Federal Savings and Loan Association of Denver, quoted in *Denver Water News* 7, no. 2 (February 1939): 3.
14. Federal Writers' Project, *Colorado: A Guide to the Highest State* (New York: Hastings House, 1941), 128; Robert L. Perkin, *Denver: The First Hundred Years* (New York: Doubleday & Company, 1959), 19.
15. Brian A. Ellison, *The Denver Water Board: Bureaucratic Power and Autonomy in Local Natural Resource Agencies* (PhD dissertation, Colorado State Univ.–Ft. Collins, 1993), 68, 78; "City Purchases Water from Western Slope," *The Denver Post,* November 16, 1932; James L. Cox, *Metropolitan Water Supply: The Denver Experience* (Boulder: Bureau of Governmental Research and Service and the University of Colorado, 1967), 94–95; "That's That," *The Denver Post,* November 8 and 16, 1932,

and March 17, 1933; "Water Report Halts Board's Secret Deal," *The Denver Post*, November 8, 1932.

16. "That's That," *The Denver Post*, November 16 and December 16 and 20, 1932, and March 17, 1933; "Water Board Amendment Most Important Issue to Be Settled," *Rocky Mountain News*, May 16, 1933; "Denver's Lusty Answer," *Rocky Mountain News*, May 17, 1933.
17. "Avalanche of Votes Kills Water Board Amendment by over 2 to 1," *Rocky Mountain News*, May 17, 1933; "Block Plot to Kill Water Board, Van Cise Exhorts," *Rocky Mountain News*, May 16, 1933; "City Requests Drastic Action to Meet Crisis," *The Denver Post*, March 16, 1933; "Petitions to Abolish Water Board Ready to Be Circulated," *The Denver Post*, April 6, 1933; "So the People May Know," *The Denver Post*, March 16, 1933; "That's That," *The Denver Post*, March 16 and 17, 1933; "Water Board Amendment Most Important Issue to Be Settled," *Rocky Mountain News*, May 16, 1933; "Water Board Pushes Plan for 1934 Diversion Project," *Rocky Mountain News*, March 18, 1933.
18. Maurice Leckenby, "State Water Famine Worst in Many Years," *Rocky Mountain News*, September 25, 1932; Walter Eha, *The Moffat Tunnel Project: An Achievement in Denver's Metropolitan Development Program* (Denver: Denver Water Board of Commissioners, 1936), 3.
19. Denver Water, *Water Supply Fact Book* 2005/2006 (Denver: Denver Water, 2006); *High Line Canal: Winding Through History* (Denver: Denver Water Office of Community Affairs, 1983); D. C. Wyatt to W. P. Miller; "Antero Case Develops New Perils to City Taxpayers," *Rocky Mountain News*, January 19, 1922; "Antero Waters Needed for Farmers Now Served, None is Left for Denver," *Rocky Mountain News*, January 24, 1922; "Antero Waterless Reservoir, City Would Lose by Deal," *Rocky Mountain News*, January 22, 1922; "Denver Water Board Closes Antero Deal," *Rocky Mountain News*, May 24, 1924.
20. *Features of the Denver Water System* (Denver: Denver Water Office of Community Affairs, 1976); "Plans for Dam in Eleven Mile Canyon on File," *The Denver Post*, April 10, 1930; Denver Water, *Water Supply Fact Book* 2005/2006 (Denver: Denver Water, 2006).
21. Donald Barnard Cole, "Transmountain Water Diversion in Colorado," Part 1, *The Colorado Magazine* 25, no. 2 (March 1948): 55–58.
22. A. Lincoln Fellows, Charles P. Allen, and Edmond C. Van Diest, "Report to the Public Utilities Commission of the City and County of Denver," January 16, 1914, 9, Denver Water Records and Document Administration; Mosley, *History of the Denver Water System*, Vol. 1, 677, 682, 690, Meeker quoted on 690.
23. Cox, *Metropolitan Water Supply*, 57–58, 1922 report quoted on 58; Earl L. Mosley, "Western Slope Water Development for Denver," *Journal of the American Water Works Association* 49 (March 1957): 251–53.
24. Cox, *Metropolitan Water Supply*, 68–71, Justice Otto Bock quoted on 69. See also James Corbridge Jr., and Teresa Rice, *Vranesh's Colorado Water Law, Revised Edition* (Boulder: Univ. Press of Colorado, 1967), 69.
25. Cox, *Metropolitan Water Supply*, 59–60; Eha, *Water for Denver*, 23–29; "Denver Goes to West Slope for Additional Water Supply," *Engineering News-Record* 115 (July–Dec 1935): 358.
26. Stephen Leonard and Thomas Noel, *Denver: Mining Camp to Metropolis* (Niwot: Univ. Press of Colorado, 1990), 209; Charles C. Fisk, *The Metro Denver Water Story: A Memoir* (Fort Collins: Colorado State Univ. Libraries, 2005), 90; "Denver Goes to West Slope for Additional Water Supply," *Engineering News-Record* 115 (July–Dec. 1935): 357; Cox, *Metropolitan Water Supply*, 61; Eha, *The Moffat Tunnel Project*, 7–8; Eha, *Water for Denver*, 7–8.
27. Walter Eha, "First Western Slope Water Was Diverted During the 'Thirties," *Denver Water News* 11, no. 2 (Feb. 1943): 2.

28. "Denver Goes to West Slope for Additional Water Supply," *Engineering News-Record* 115 (July–Dec. 1935): 357; "Denver Water Supply Falling," *Rocky Mountain News*, November 6, 1932; Eha, *The Moffat Tunnel Project*, 9, 13–21; Eha, *Water for Denver*, 34.
29. George Cranmer, Letter to Stuart P. Cooke, October 27, 1939, located in George E. Cranmer Papers, WH479, Box 2, FF22, Western History Collection, Denver Public Library; Bill Miller, "Vegetable Embargo Brought Change," *Rocky Mountain News*, June 27, 1959; Paul J. Connor, "Plant to Prevent River of Filth Diseases," *Rocky Mountain News*, November 11, 1937; "Williams Fork Plan Expected to Make Money," *The Denver Post*, May 7, 1936; "Denver to Build Three-Mile Water Tunnel to Aid Sewage," *The Denver Post*, December 3, 1936.
30. Denver Water Department, *Features of the Denver Water System* (Denver: Board of Water Commissioners, 1976); Cox, *Metropolitan Water Supply*, 59–63; Earl L. Mosley, "Western Slope Water Development for Denver," *Journal of the American Water Works Association* (March 1957): 258–60.
31. James L. Cox, *Metropolitan Water Supply: The Denver Experience* (Boulder: Bureau of Governmental Research and Service and the University of Colorado, 1967), 70–71; Gregory J. Hobbs, "Green Mountain Reservoir: Lock or Key?" *Colorado Water Rights* 3, no. 2 (Summer 1984); Daniel Tyler, *Last Water Hole in the West: The Colorado–Big Thompson Project and the Northern Colorado Water Conservancy District* (Niwot: Univ. of Colorado Press, 1992), 72–73; "Diversion Project Heads for Trouble," *Rocky Mountain News*, March 4, 1936.

Chapter Five

A Horrifying Jigsaw Puzzle

1. Gary Harmon, "Denver Water Manager Says It's No Longer 'Evil Empire,'" *Grand Junction Daily Sentinel*, September 17, 2010.
2. Arthur Meier Schlesinger, *The Rise of the City, 1878–1898* (New York: Macmillan Company, 1933), 86–87.
3. Donald Worster, *Rivers of Empire: Water, Aridity, and the Growth of the American West* (New York: Pantheon, 1985), 52, 57.
4. William Cronon, *Nature's Metropolis: Chicago and the Great West* (New York: W.W. Norton, 1991).
5. Stephen Leonard and Thomas Noel, *Denver: Mining Camp to Metropolis* (Niwot: Univ. Press of Colorado, 1990), 116, xi, 44; Kathleen A. Brosnan, *Uniting Mountain and Plain: Cities, Law, and Environmental Change along the Front Range* (Albuquerque: Univ. of New Mexico Press, 2002) 7–8.
6. Leonard and Noel, *Denver*, 119, 123; Brosnan, *Uniting Mountain and Plain*, 197.
7. Leonard and Noel, *Denver*, 235, 250.
8. Denver Board of Water Commissioners Meeting Minutes from August 23, 1951.
9. Brian A. Ellison, *The Denver Water Board: Bureaucratic Power and Autonomy in Local Natural Resource Agencies* (PhD dissertation, Colorado State Univ.–Ft. Collins, 1993), 142–48; J. Gordon Milliken, "Water Management Issues in the Denver, Colorado, Urban Area," in *Water and Arid Lands of the Western United States*, eds., Mohamed T. El-Ashry and Diana C. Gibbons (Cambridge, MA: Cambridge Univ. Press, 1988), 339–41; James L. Cox, *Metropolitan Water Supply: The Denver Experience* (Boulder: Bureau of Governmental Research and Service and the University of Colorado, 1967), 155.
10. Denver Board of Water Commissioners Meeting Minutes from March 26, 1948; Cox, *Metropolitan Water Supply*, 111, 123–24.
11. *Charter of the City and County of Denver, Adopted by a Vote of the People March 29, 1904, with all Amendments Thereof to and including January 1st, 1927,* Article XIX, Sections 264-C and 264-D.
12. Cox, *Metropolitan Water Supply*, 124, 126.

13. Cox, *Metropolitan Water Supply*, 14–28.
14. Leonard and Noel, *Denver*, xiii, 277.
15. Ruth McManus and Philip J. Ethington, "Suburbs in Transition: New Approaches to Suburban History," *Urban History* 34, no. 2 (2007), 319; Jon C. Teaford, *The Twentieth-Century American City*, 2nd ed. (Baltimore: Johns Hopkins Univ. Press, 1993); Martin V. Melosi, *The Sanitary City: Urban Infrastructure in America from Colonial Times to the Present* (Baltimore: Johns Hopkins Univ. Press, 2000), 207.
16. Thomas B. McKee, et. al., *A History of Drought in Colorado: Lessons Learned and What Lies Ahead* (Fort Collins: Colorado Water Resources Research Institute, 2000), 15.
17. Letter from Robert Millar, DWB Manager, to Board of Water Commissioners, December 7, 1959.
18. J. Gordon Milliken, "Water Management Issues," 340.
19. Charles C. Fisk, *The Metro Denver Water Story: A Memoir* (Fort Collins: Colorado State Univ. Libraries, 2005), 81–82, 111; Clyde King, *The History of the Government of Denver with Special Reference to its Relations with Public Service Corporations* (Denver: Fisher Book Company, 1911), 222, 225–28; Cox, *Metropolitan Water Supply*, 5–8, 49–50.
20. J. Gordon Milliken, "Water Management Issues," 339; Cox, *Metropolitan Water Supply*, 104–5.
21. J. Gordon Milliken, "Water Management Issues," 345; George Creamer quoted in Fisk, *Metro Denver Water Story*, 113.
22. In 1963, the population served by Denver Water was 649,000 and the average consumption per day was 144,086,356 gallons. Denver Board of Water Commissioners, *Annual Statistical Report of the Board of Water Commissioners, City and County of Denver, Year Ending December 31, 1963* (Denver Board of Water Commissioners), 14. In 2000, the population served was 1,036,000 people and the average daily consumption was 228,380,000 gallons per day. Denver Water, *Comprehensive Annual Financial Report: For the Year Ended December 31, 2006* (Denver: Denver Water Department, 2007), iii–75. The prediction and quotation appear in Board of Water Commissioners, *Report of the Board of Water Commissioners, Year Ending December 31, 1953*, 7–8; Mosley quoted in Nello Cassai, "Water Edict Dooms 6,000 Homes: End of City's Expansion Seen by '63," *The Denver Post*, October 19, 1954.
23. Cox, *Metropolitan Water Supply*, 73–74; Board of Water Commissioners, *Report of the Board of Water Commissioners: Year Ending December 31, 1955* (Denver, CO: Board of Water Commissioners, 1956), 6; Mosley, "Western Slope Water Development for Denver," 256–57; Board of Water Commissioners, *Report of the Board of Water Commissioners: Year Ending December 31, 1954* (Denver: Board of Water Commissioners, 1955), 8, 50.
24. Alvord, Burdick & Howson, *Report on Future Water Supply, Denver Municipal Water Works, Denver, Colorado, March 1955* (Chicago: Alvord, Burdick & Howson, 1955); Board of Water Commissioners, *Report of the Board of Water Commissioners, Year Ending December 31, 1954* (Denver: Board of Water Commissioners, 1955), 8; Daniel Tyler, *Last Water Hole in the West: The Colorado–Big Thompson Project and the Northern Colorado Water Conservancy District* (Niwot: Univ. of Colorado Press, 1992), 209.
25. John W. Buchanan, "Use of Blue River Up to Courts," *The Denver Post* (July 30, 1950). This is one of many reports that note the significance of the gap in the Water Board's attention to the Blue River.
26. Board of Water Commissioners, *Report of the Board of Water Commissioners: Year Ending December 31, 1952* (Denver: Board of Water Commissioners, 1953), 8; Fisk, *Metro Denver Water Story*, 95; Tyler, *Last Water Hole*, 163.
27. *The City of Grand Junction v. The City and County of Denver*, 1998 Colorado Supreme Court No. 97SA93; Gregory J. Hobbs Jr., "State Water Politics Versus an Independent Judiciary: The Colorado and Idaho Experiences," *Quinnipiac Law Review* 20 (2001): 679–83; Tyler, *Last Water Hole*, 163–64; Cox, *Metropolitan Water Supply*, 71–81.

28. Cox, *Metropolitan Water Supply*, 69.
29. Cox, *Metropolitan Water Supply*, 71–72.
30. "City Failed to Establish Claim, Jurist Says," *The Denver Post*, October 19, 1954; Tyler, *Last Water Hole*, 163, 192, 207–8; Cox, *Metropolitan Water Supply*, 72–73; Nello Cassai, "Reported Court Ruling Leaves City Dilemma," *The Denver Post*, October 18, 1954.
31. "Denver's Water Crisis Not Over, Expert Warns," *The Denver Post*, October 1, 1954.
32. "City Failed to Establish Claim, Jurist Says," *The Denver Post*, October 19, 1954; Colorado Water Conservation Board, *The Green Mountain Reservoir Problem* (Denver: Colorado Water Conservation Board, 1960), 203; Tyler, *Last Water Hole*, 208; Gregory J. Hobbs Jr., "State Water Politics Versus an Independent Judiciary," 681; Nello Cassai, "Reported Court Ruling Leaves City Dilemma," *The Denver Post*, October 18, 1954; "The Supreme Court vs. a Growing City," *The Denver Post*, October 20, 1954.
33. "Court Kills Denver Hope for Water," *Rocky Mountain News*, October 19, 1954; Tom Gavin, "Denver's Hope for Water Is Killed," *Rocky Mountain News*, October 19, 1954; "Split Court Blocks Water Hope," *Rocky Mountain News*, October 20, 1954.
34. Nello Cassai, "End of City Expansion Seen by '63," *The Denver Post*, October 19, 1954; "The Supreme Court vs. a Growing City," *The Denver Post*, October 20, 1954; Ed Oschmann, "Denver Board Likely to Halt All Irrigation," *Rocky Mountain News*, October 26, 1954; Nello Cassai, "All Sprinkling Banned: City Puts Ban on Use of Hoses," *The Denver Post*, October 26, 1954.
35. "Water Decision Stands," *Grand Junction Sentinel*, December 13, 1954; Tom Gavin, "Denver Keeps Alive Hopes for Blue River," *Rocky Mountain News*, December 14, 1954.
36. Tyler, *Last Water Hole*, 192–93, 205–15; Cox, *Metropolitan Water Supply*, 71–85; Nello Cassai, "Blue River Water Feud Seen Set for Peaceful Settlement," *The Denver Post*, October 4, 1955.
37. Judge William Lee Knous, "Final Decree" regarding Consolidated Cases Civil Nos. 2782, 5016, and 5017 (October 12, 1955); Cox, *Metropolitan Water Supply*, 72–73; Tyler, *Last Water Hole*, 213; "Denver's Water Crisis Not Over, Expert Warns," *The Denver Post*, October 1, 1954.
38. Tyler, *Last Water Hole*, 214–15; David Rose, "Dillon Dam Water Conflict Ends in Plan Agreement," *Rocky Mountain News*, April 17, 1964; Cox, *Metropolitan Water Supply*, 74–79; Nello Cassai, "City Won't Get Blue Water for at Least Five Years," *The Denver Post*, October 6, 1955.
39. Nello Cassai, "City Won't Get Blue Water for At Least Five Years," *The Denver Post*, October 6, 1955.
40. Earl L. Mosley, *Blue River Water Diversion Project: Design and Construction Features*, vol. 1 (Denver: Board of Commissioners, City and County of Denver, 1962) provides an overview of the diversion project as it neared completion; Cox, *Metropolitan Water Supply*, 67; Mary Ellen Gilliland, *Summit: A Gold Rush History of Summit County, Colorado* (Silverthorne, CO: Alpenrose, 1980), 79–91; Reva Cullen, "Saddened Dillon Residents Watch Roberts Tunnel Ceremonies," *Rocky Mountain News*, July 22, 1956; Sandra F. Mather, *Summit County* (Charleston, SC: Arcadia, 2008); Sandra F. Pritchard, *Dillon, Denver, and the Dam* (Dillon, CO: Summit Historical Society, 1994).
41. Lee Olson, "City's New Water Plan Draws West Slope Ire," *The Denver Post*, October 12, 1960; Cox, *Metropolitan Water Supply*, 72–82; Tyler, *Last Water Hole*, 295–97, 308–9 (Saunders quotation on 308).
42. Bill Kostka Jr., "Denver's Claim to Green Water Rebuffed by US," *Rocky Mountain News*, August 22, 1959; "US to Study Eastern Slope Plea for Green Mountain Water," *The Denver Post*, November 19, 1959; Gene Wortsman, "Interior Department to Study Denver Water's Claims," *Rocky Mountain News*, November 19, 1959; Barnard quoted in Tyler, *Last Water Hole*, 309.

43. "Denver Presses Claim to Green Mountain Water," *The Denver Post,* March 11, 1960; Lee Olson, "City's New Water Plan Draws West Slope Ire," *The Denver Post,* October 12, 1960; "Inter-Slope Water Warfare Erupts Again," *Rocky Mountain News,* October 12, 1960.
44. Cox, *Metropolitan Water Supply,* 73–74.
45. Cox, *Metropolitan Water Supply,* 73; Ellison, *Denver Water Board,* 95.
46. David Rose, "Others Join to Fight City Over Dillon Dam," *Rocky Mountain News,* September 12, 1963; Cox, *Metropolitan Water Supply,* 80–82; Tyler, *Last Water Hole,* 310.
47. Mosley, *Blue River Water Diversion Project,* 5; Cox, *Metropolitan Water Supply,* 66–67; Reva Cullen, "Saddened Dillon Residents Watch Roberts Tunnel Ceremonies," *Rocky Mountain News,* July 22, 1956; Sandra F. Pritchard, *Dillon, Denver and the Dam,* 9, 26, 28–29, 42; Mary Ellen Gilliland, *Summit,* 79–91.
48. Bert Hanna, "Attorney Says Denver Conforming with Law," *The Denver Post,* September 2, 1963; David Rose, "Others Join to Fight City Over Dillon Dam," *Rocky Mountain News,* September 12, 1963; Bert Hanna, "Legal Threats Posed to Denver Water Diversion Plan," *The Denver Post,* August 11, 1963.
49. Bert Hanna, "US Demands Denver Halt Dillon Dam Work," *The Denver Post,* September 1, 1963; Bert Hanna, "Dillon Dam Plug Job Completed," *The Denver Post,* September 6, 1963; Bert Hanna, "Storage at Dillon Dam Starts Despite Dispute," *The Denver Post,* September 3, 1963; Bert Hanna, "Trial Delay Raises Hopes of Out-of-Court Dillon Settlement," *The Denver Post,* January 19, 1964; Tyler, *Last Water Hole,* 295–97, 312–16; David Rose, "Others Join to Fight City Over Dillon Dam," *Rocky Mountain News,* September 12, 1963; Cox, *Metropolitan Water Supply,* 83–84.
50. James O. Wood and Gene Cooper, "Denver Defies US Over Dillon Dam," *Rocky Mountain News,* September 2, 1963.
51. Bert Hanna, "Storage at Dillon Dam Starts Despite Dispute," *The Denver Post,* September 3, 1963; Del W. Harding, "Denver Begins Storing Water in Dillon Dam," *Rocky Mountain News,* September 4, 1963; "Denver Plugs Up Blue to Begin Dam Storage, *Grand Junction Sentinel,* September 3, 1963.
52. David Rose and Del W. Harding, "US Suit Challenges Denver's Dillon Rights," *Rocky Mountain News,* September 7, 1963; Bert Hanna, "Denver Faces Court Fight in Dillon Dispute," *The Denver Post,* September 8, 1963; David Rose, "Others Join to Fight City Over Dillon Dam," *Rocky Mountain News,* September 12, 1963.
53. *Grand Junction Sentinel,* September 13, 1963, quoted in Tyler, *Last Water Hole,* 313; Editorial, "Eloquent Appeal," *Grand Junction Sentinel,* October 18, 1963; "Water Could Make Currigan New Style Political Hero," *Denver Democrat,* October 26, 1963; William H. Nelson, "Water Negotiations Fail; Trial Preparations Continue," *Grand Junction Sentinel,* October 27, 1963.
54. Contemporary news reports pegged the total cost at between $70 million and $77 million. See Bert Hanna, "Trial Delay Raises Hopes of Out-of-Court Dillon Settlement," *The Denver Post,* January 19, 1964, 32C; William Logan, "Dillon Water Pours From Roberts Tunnel," *Rocky Mountain News,* July 18, 1964, 22.
55. *Colorado River Compact,* 1922, Article IV, Section (b).
56. David Rose, "Denver Offers Plan in Dillon Dam Fight," *Rocky Mountain News,* September 22, 1963; Tom Hutton, "City Offers Compromise," *The Denver Post,* September 27, 1963. Tyler, *Last Water Hole,* 314–16.
57. "Saunders Asks Parley in Dillon Dam Dispute," *The Denver Post,* September 9, 1963; Del W. Harding, "Conference Urged on Dillon Reservoir," *The Denver Post,* September 10, 1963; "Meeting Sought in Dillon Dam Dispute," *Rocky Mountain News,* September 14, 1963; "Delay in Dillon Case Termed Not Harmful," *The Denver Post,* December 4, 1963; Bert Hanna, "Dillon Bid Pleases Saunders," *The Denver Post,* December 4, 1963; "Trial Delay Raises Hopes of Out-of-Court Dillon Settlement," *The Denver Post,* January 19, 1964.

58. "Denver Water Board OK's Blue River Pact," *The Denver Post*, April 11, 1964; "Peace Hopes Revived in Blue River Battle," *The Denver Post*, April 14, 1964; Tyler, *Last Water Hole*, 315–16.
59. David Rose, "Dillon Dam Water Conflict Ends in Plan Agreement," *Rocky Mountain News*, April 17, 1964, Tyler, *Last Water Hole*, 315–16.
60. Tyler, *Last Water Hole*, 314–16; Cox, *Metropolitan Water Supply*, 84; William Logan, "Dillon Water Pours from Roberts Tunnel," *Rocky Mountain News*, July 18, 1964.
61. William Logan, "Dillon Water Pours from Roberts Tunnel," July 18, 1964.
62. Ellison, *Denver Water Board*, 38.
63. Jonathan Swift, *Gulliver's Travels* (1726); reprinted edition edited by Peter Dixon and John Chalker (Harmondsworth, England: Penguin, 1975), 55–57.
64. Black & Veatch, *Report on Comprehensive Water System Study*, prepared for the Board of Water Commissioners, City and County of Denver by Black & Veatch, Consulting Engineers, Kansas City, Missouri (1963), 32, quoted in Cox, *Metropolitan Water Supply*, 89, full source citation at fn. 7, p. 59.

Chapter Six

No Country for Old Habits

1. Mark Twain, *Tom Sawyer Abroad* (1878; reprinted New York, Harper and Brothers, 1910), 77.
2. Hugh Bowden, *Classical Athens and the Delphic Oracle: Divination and Democracy* (Cambridge, UK: Cambridge Univ. Press, 2004), 12–39. See also H. W. Parke, *A History of the Delphic Oracle* (Oxford, UK: Basil Blackwell, 1939).
3. Board of Water Commissioners, *Report of the Board of Water Commissioners: Year Ending December 31, 1951* (Denver: Board of Water Commissioners, 1952), 6–7; Board of Water Commissioners, *Report of the Board of Water Commissioners: Year Ending December 31, 1955* (Denver: Board of Water Commissioners, 1956), 6. Cox, *Metropolitan Water Supply*, 63.
4. "The Day the Sky Turned Black," *The Denver Post Empire Magazine*, November 9, 1958.
5. Mark Harvey, *A Symbol of Wilderness: Echo Park and the American Conservation Movement*, Weyerhaeuser Environmental Classic edition (Seattle: Univ. of Washington Press, 2000), xix–xx.
6. Harvey, *A Symbol of Wilderness*, 297; William Cronon, foreword to Harvey, *A Symbol of Wilderness*, xiii.
7. Harvey, *A Symbol of Wilderness*, xvii.
8. Harvey, *A Symbol of Wilderness*, xxii, xvii.
9. Denver Water, "Water Supply Fact Book" (Denver: Denver Water, 2005), 12.
10. Alex Shoumatoff, "The Skipper and the Dam," *The New Yorker*, December 1, 1986. Denver Water Board Planning Director Robert Taylor explained that "Two Forks would enable us to collect high flows wherever they occur, including West Slope water released through Roberts Tunnel into the North Fork."
11. "Cranmer Urges Two Forks High Dam," *The Denver Post*, November 16, 1965; Daniel F. Luecke, "Two Forks Dam and Endangered Species," *Colorado Water* 26, no. 2 (March/April 2009): 17.
12. D. D. Gross, Chief Engineer of the Board of Water Commissioners, "Two Forks Reservoir: Tabulation of Historical Information," June 23, 1954; Bert Hanna, "Bureau Report Favors 2 South Platte Dams," *The Denver Post*, March 13, 1966; Dick Prouty, "Two Forks Dam Stirs Debate," *The Denver Post*, June 14, 1970; Dick Prouty, "Two Forks Dam Plan Challenged," *The Denver Post*, December 17, 1970.
13. Taylor quoted in Shoumatoff, "The Skipper and the Dam."
14. Denver Water, "2010 Water Quality Report."
15. Steve Hinchman, "EPA to Denver: Wake Up and Smell the Coffee!" in Char Miller,

ed., *Water in the West: A High Country News Reader* (Corvallis: Oregon State Univ. Press, 2000), 204.

16. "Water Board Counsel Hits Wilderness Plan," *Rocky Mountain News*, November 22, 1958.
17. "Denver Water Claims Hit at Hearing," *The Denver Post*, November 20, 1959.
18. 1974 Colorado Session Laws, "Areas and Activities of State Interest," Chapter 80, 335–55, now codified at CRS §§24-65.1-101 to -502; *City and Country of Denver, acting by and through its Board of Water Commissioners v. Board of County Commissioners of Grand County*, 782 P.2d 753, Supreme Court of Colorado, Nov. 13, 1989.
19. The water system first exceeded its capacity of 450 million gallons per day on July 13 and 14, 1972, when Denver residents drew 461.1 and 463.9 gallons through the pipes, respectively. The record-breaking July 6, 1973, usage was the "high point" of a hot summer in which the Water Board exceeded the system capacity five times. Denver Board of Water Commissioners, Annual Statistical Report Year Ending December 31, 1972 (Denver: Board of Water Commissioners, 1973); Denver Board of Water Commissioners, Annual Statistical Report Year Ending December 31, 1973 (Denver: Board of Water Commissioners, 1974).
20. Bob Jain, "Voters Reject Water Board Plan by 3,754 Margin," *The Denver Post*, July 12, 1972; "Water Issue Foes Desire 'Another Plan,'" *Rocky Mountain News*, July 13, 1972; "Denver's Water Dilemma: How Foothills Became Stalled," editorial, *The Denver Post*, December 4, 1977; "Denver's Water Dilemma: Arguments Against Foothills," editorial, *The Denver Post*, December 4, 1977.
21. Bob Ewegen, "Water Bond Issue Passes," *The Denver Post*, November 7, 1973; Bob Ewegen and Andy Rogers, "Study to Weigh Whether Bonds Set New Course," *The Denver Post*, November 7, 1973; "Arguments Against Foothills," *The Denver Post*; Charles C. Fisk, *The Metro Denver Water Story: A Memoir* (Fort Collins:, Colorado State Univ. Libraries, 2005), 118–22; Brian A. Ellison, "Denver Water Politics, Two Forks, and the Implication for Development on the Great Plains," *Water on the Great Plains: Issues and Policies*, eds. Peter J. Longo and David W. Yoskowitz (Lubbock: Texas Tech Univ. Press, 2002), 102–4; J. Gordon Milliken, "Water Management Issues in the Denver, Colorado, Urban Area," *Water and Arid Lands of the Western United States*, eds., Mohamed T. El-Ashry and Diana C. Gibbons (Cambridge, MA: Cambridge Univ. Press, 1988), 353; Lawrence S. Bacow and Michael Wheeler, *Environmental Dispute Resolution* (New York: Plenum Press, 1984), 197.
22. Bacow and Wheeler, *Environmental Dispute Resolution*, 199.
23. "How Foothills Became Stalled," *The Denver Post*; Brian A. Ellison, *The Denver Water Board: Bureaucratic Power and Autonomy in Local Natural Resource Agencies* (PhD dissertation, Colorado State Univ.–Ft. Collins, 1993), 105; Suzanne Weiss, "Regional EPA Chief Urges Foothills' Demise," *Rocky Mountain News*, September 27, 1977; Al Gordon, "EPA Official Wants 'Less Expendable Design': Merson Seeks Future Foothills Limits," *Rocky Mountain News*, April 22, 1978; Al Gordon, "Water Board Challenges EPA Permit Conditions," *Rocky Mountain News*, April 29, 1978; Bacow and Wheeler, *Environmental Dispute Resolution*, 199–200.
24. "Merson Rejects Foothills Permit," *Rocky Mountain News*, May 26, 1978; Bacow and Wheeler, *Environmental Dispute Resolution*, 200–201.
25. Bacow and Wheeler, *Environmental Dispute Resolution*, 201.
26. John Muir, *My First Summer in the Sierra* (Boston: Houghton Mifflin, 1911), 211.
27. Bacow and Wheeler, *Environmental Dispute Resolution*, 204.
28. Quoted in Bacow and Wheeler, *Environmental Dispute Resolution*, 211.
29. Whit Sibley, "Board Seeking $36 Million in Foothills Suit," *The Denver Post*, August 22, 1978; Bacow and Wheeler, *Environmental Dispute Resolution*, 225–26.
30. Jerry Brown, "Defendant Calls Water Suit a Bid to Limit Speech," *Rocky Mountain News*, January 17, 1979; Bacow and Wheeler, *Environmental Dispute Resolution*, 233–34.

31. Denver Water Board, *The Foothills Complex: A Commitment to the Future* (Denver: Denver Water Board, 1984); W. H. Miller, "Mandatory Water Conservation and Tap Allocations in Denver, Colorado," *Journal of the American Water Works Association*, 70, No. 2 (1978), 60–63; Ellison, *Denver Water Board*, 106.
32. "Merson Rejects Foothills Permit," *Rocky Mountain News*, May 26, 1978; Jerry Brown, "Foothills Backers Still Hold Hope," *Rocky Mountain News*, May 26, 1978; "Water Officials Reject Merson Suggestions," *Rocky Mountain News*, May 27, 1978; Bill Pardue, "Foothills First Steps Taken," *The Denver Post*, May 28, 1978; Denver Water Board, *The Foothills Complex.*
33. Bacow and Wheeler, *Environmental Dispute Resolution*, 214.
34. Timothy Wirth et al., "Principles of Agreement" (January 3, 1979) and James B. Kenney Jr., et al., "Memorandum of Understanding: Denver Water Board and Environmental Defendants in Denver v. Andrus"; Bacow and Wheeler, *Environmental Dispute Resolution*, 232–36.
35. Frances Melrose, "Water Officials OK 1,800 Extra Taps," *Rocky Mountain News*, June 21, 1979; Denver Water Board, *The Foothills Complex.*
36. Ellison, *Denver Water Board*, 115; "Haskell's Wilderness Bill Gaining Momentum," *San Miguel Forum*, May 28, 1975; "Eagles Nest Wilderness Area in Colorado Approved," *Ellensburg Daily Record*, June 22, 1976; Suzanne Weiss, "Water for 1 Million at Stake," *The Denver Post*, August 17, 1975; Suzanne Weiss, "Western Slope Water Rights Denied," *The Denver Post*, September 28, 1977; "Water Board Buys Williams Fork Rights," *Rocky Mountain News*, February 27, 1963; "Water Board Faces New Fight," *The Denver Post*, January 9, 1977; "Water Board Skips Bids, Risks Millions," *The Denver Post*, December 31, 1978; "Gamble Challenged; US Halts Williams Fork Work," *The Denver Post*, January 17, 1979; "Water Board Sues over Williams Fork," *Rocky Mountain News*, May 24, 1979; "Feud Over Williams Fork project pits Water Board, Forest Service," *Rocky Mountain News*, July 16, 1979; "Judge Denies Request to Finish Water Project," *Rocky Mountain News*, August 4, 1979; "Kane Won't Intercede in Williams Fork Dispute," *Rocky Mountain News*, January 16, 1980; "City's Williams Fork Project Stalled By Court," *Rocky Mountain News*, June 2, 1981; "Right-of-way Ruling Fought by Water Unit," *Rocky Mountain News*, June 16, 1981; "Court Upholds Water Diversion Halt," *Rocky Mountain News*, December 10, 1982; Denver Water Department, *1987 Annual Report: Year Ending December 31, 1987* (Denver Water Department, 1988), 14. As a measure of their stubbornness, Denver Water's 1987 annual report includes a map that shows the Eagle–Piney, Eagle–Colorado, and East Gore projects as "under development."
37. Dan Luecke, "Two Forks: The Rise and Fall of a Dam," *Natural Resources and Environment* 14 (Summer 1999): 25–26; Susan L. Carpenter and W. J. D. Kennedy, "The Denver Metropolitan Water Roundtable: A Case Study in Researching Agreements," *Natural Resources Journal* 28 (Winter 1988): 21–35. The members of the Colorado Environmental Caucus are listed in Fisk, "Metro Denver Water Story," 126, and Colorado Environmental Caucus, "Metropolitan Denver's Future Water Supply."
38. Luecke, "Two Forks," 26; Ed Marston, "Water Pressure," *High Country News*, November 20, 2000.
39. Fisk, "Metro Denver Water Story," 126; Luecke, "Two Forks," 26; Carpenter and Kennedy, "Denver Metropolitan Water Roundtable," 21–35.
40. "West Slope, Denver Cease Firing in War Over Two Forks Proposal," *Rocky Mountain News*, December 4, 1986; Janet Day, "Denver Lines Up Support for Two Forks," *Rocky Mountain News*, December 9, 1986; Janet Day, "N. Coloradans Back Two Forks," *Rocky Mountain News*, April 28, 1988.
41. Information on the chain of coalitions and agreements formed in the early 1980s around Two Forks can be found in Denver Water Department, *1983 Annual Report: Year Ending December 31, 1983* (Denver: Denver Water Department, 1984); Denver

Water Department, *1984 Annual Report; Year Ending December 31, 1984* (Denver: Denver Water Department, 1985); Denver Water Planning and Finance Divisions, *Annual Report and Component Unit Financial Report for the Year Ended December 31, 1988* (Denver: Denver Water, 1989); Daniel Luecke, "Two Forks: The Rise and Fall of a Dam," *Natural Resources and Environment* 14 (1999), 24–28, 69.

42. Brian A. Ellison, "Denver Water Politics," 101.
43. Brian A. Ellison, *The Denver Water Board,* 109–13; Franklin J. James and Christopher B. Gerboth, "A Camp Divided: Annexation Battles, the Poundstone Amendment, and Their Impact on Metropolitan Denver, 1941–1988," *Colorado History* 5 (2001): 144.
44. Kenneth T. Jackson, *The Crabgrass Frontier: The Suburbanization of the United States* (New York: Oxford Univ. Press, 1985), 289; Teaford, *The Twentieth Century American City,* 2nd ed. (Baltimore: Johns Hopkins Univ. Press, 1993), 109.
45. Frederick D. Watson, "Removing the Barricades from the Northern Schoolhouse Door: School Desegregration in Denver" (PhD Dissertation, Univ. of Colorado–Boulder, 1993), i–ii, 176.
46. Watson, "Removing the Barricades," 213; Gerboth and James, "A Camp Divided," 131.
47. Teaford, *The Twentieth Century American City,*109.
48. Ellison, "The Denver Water Board," 111–13.
49. Janet Day, "Dam to Ruin Crane Habitat, Audubon Warns," *Rocky Mountain News,* March 26, 1988; Craig, Gerald R. and James H. Enderson, *Peregrine Falcon Biology and Management in Colorado, 1973–2001* (Denver: Colorado Division of Wildlife, 2004); Luecke, "Two Forks Dam and Endangered Species," 18.
50. "Do Butterflies Swim? Two Forks a Cosmic Question for Rare Skipper," *The Denver Post,* April 20, 1985; Donald Myers, "In Danger: The Pawnee Montana Skipper: Is Its Future Worth a Dam?" *Rocky Mountain News Sunday Magazine,* July 14, 1985. Alex Shoumatoff, "The Skipper and the Dam."
51. Luecke, "Two Forks," 26–27; Ellison, *Denver Water Board,* 121–22; "Amendment to Memorandum of Agreement of November 2, 1982," Denver Water Board minutes, December 17, 1985; "Two Forks Dam and Reservoir: A Historical Look," (March 1, 1990), located in Box 185001938, Denver Water Records and Document Administration.
52. Denver Water Department, *1987 Annual Report: Year Ending December 31, 1987* (Denver Water Department, 1988); Denver Water Planning and Finance Divisions, *Annual Report and Component Unit Financial Report for the Year Ended December 31, 1988* (Denver: Denver Water, 1989); Luecke, "Two Forks," 27; Shoumatoff, "The Skipper and the Dam," 12; Ellison, "Denver Water Politics," 107; James Coates, "Impact Statement on Dam Becomes a Great Big Business," *Chicago Tribune,* July 27, 1986.
53. Daniel F. Luecke, "Two Forks: The Rise and Fall of a Dam," *Natural Resources and Environment,* 14. No. 24 (1999), 24–28, 69; Mark Obmasik, "Two Forks 'Done Deal' to EPA Officials: Ex-Project Chief Contends Agency Higher-Ups Asked Staff to Justify," *The Denver Post,* October 11, 1989.
54. Environmental Protection Agency, "Clean Water Act: Section 404(c) 'Veto Authority,'" (Washington, DC: Environmental Protection Agency, 2010). Two Forks was the eleventh 404(c) Veto issued in the agency's history.
55. Philip Shenon, "E.P.A. Will Review Dam in Colorado," *The New York Times,* March 25, 1989; Michael Wesskopf, "EPA Chief Blocks Dam's Approval," *The Washington Post,* March 25, 1989; Mark Obmascik, "2 Forks Foes Used Effective Strategy," *The Denver Post,* March 26, 1989; Keith Schneider, "Colorado Dam Would Ruin Wildlife Area, an E.P.A. Official Says," *The New York Times,* August 30, 1989; William K. Reilly, "Statement of William K. Reilly on the Two Forks Dam and Reservoir," appended to a letter from James J. Scherer to Col. Steven G. West, March 24, 1989; Clean Water Act of 1972, U.S.C. § 1334.

56. Lee A. DeHihns, III, for the US Environmental Protection Agency, "Proposed Determination to Prohibit, Restrict, or Deny the Specification, or use for Specification, of an Area as a Disposal Site; South Platte River," *Federal Register* 54, no. 170 (September 5, 1989): 36862–71, quotes on 36866 and 36870.
57. "Testimony of Mayor Frederico Pena at EPA Hearings on Two Forks Dam," October 24, 1989, located in "DWD/Metro Water Providers Participants and Comments on Proposed Veto of Two Forks," Vol. IV–V, Box 179942744, Denver Water Records and Document Administration; letter from Jack F. Ross to W. H. Miller, November 17, 1989, located in "DWD/Metro Water Providers Participants and Comments on Proposed Veto of Two Forks," Vol. IV–V, Box 179942744, Denver Water Records and Document Administration; "Metro Denver Water Authority Requests Investigation of EPA," September 27, 1989, located in "DWD/Metro Water Providers Participants and Comments on Proposed Veto of Two Forks," Vol. IV–V, Box 179942744, Denver Water Records and Document Administration; letter from William L. Armstrong to Charles A. Bowsher, October 16, 1989, located in "DWD/Metro Water Providers Participants and Comments on Proposed Veto of Two Forks," Vol. IV–V, Box 179942744, Denver Water Records and Document Administration. After the public comment period, DeHihns reiterated most of his findings, often verbatim, in a subsequent report to Reilly, *Recommended Determination to Prohibit Construction of Two Forks Dam and Reservoir Pursuant to Section 404(c) of the Clean Water Act* (Washington, DC: US Environmental Protection Agency, March 26, 1990). The veto was formalized in the *Final Determination of the US Environmental Protection Agency's Assistant Administrator for Water Pursuant to Section 404(c) of the Clean Water Act Concerning the Two Forks Water Supply Impoundments, Jefferson and Douglas Counties, Colorado, November 23, 1990*; EPA, "Clean Water Act: Section 404(c) 'Veto Authority.'"
58. William K. Reilly, "William K. Reilly: Oral History Interview," conducted over multiple sessions by Dennis Williams in 1993, (Washington, DC: US Environmental Protection Agency, 1995).
59. William K. Reilly, ed., for the Task Force on Land Use and Urban Growth, *The Use of Land: A Citizen's Policy Guide to Urban Growth* (New York: Thomas Y. Crowell, 1973); Reilly, "Oral History."
60. Russell, Karsh & Hagan Public Relations, "Why Two Forks?" (no date), located in Box 185001938, Denver Water Records and Document Administration; Dan Luecke, personal interview by Patty Limerick, Jason L. Hanson, and Dylan Eiler, October 19, 2009, Center of the American West at the Univ. of Colorado at Boulder; quotations from Miller, *Water in the West*, 33, 204.
61. Ellison, *Denver Water Board*, 123–24; Fisk, "Metro Denver Water Story," 128.
62. Ellison, *Denver Water Board*, 123–24.
63. Russell, Karsh & Hagan, "Why Two Forks?"
64. "Environmentalists' Plan on Reservoir Unveiled," *Rocky Mountain News*, October 9, 1982; Luecke, personal interview.
65. "Metropolitan Denver's Future Water Supply: The Colorado Environmental Caucus's Environmentally and Fiscally Sound Alternative to the Denver Water Board's Application for a 25-Year Permit to Build Two Forks Dam" (April 1988), 1, 7, located in Box 185001938, Denver Water Records and Document Administration.
66. "Metropolitan Denver's Future Water Supply," 3. For another study of proposed alternatives with a focus on conservation and efficiency, see also John C. Woodwell, *Supplying Denver With Water Efficiency: An Alternative to Two Forks Dam* (Snowmass, CO: Rocky Mountain Institute, 1989); this report was sent to William Reilly in November 1989 by Amory Lovins, RMI's director of research (letter from Lovins to Reilly, November 22, 1989, located in "DWD/Metro Water Providers Participants and Comments on Proposed Veto of Two Forks," Vol. IV–V, Box 179942744).
67. "Metropolitan Denver's Future Water Supply," 7.

68. Suzanne Weiss, "Two Forks' Apparent Demise Damming Metro Cooperation," *Rocky Mountain News,* October 9, 1989; Joseph B. Verrengia, "Dam Backers Quitting the Fight," *Rocky Mountain News,* November 24, 1990; Dick Foster, "Two Forks Death May Trigger Water Wars," *Rocky Mountain News,* December 2, 1990; Associated Press, "Water Department Boss Discourages New Two Forks Suit," *Rocky Mountain News,* January 21, 1991; Bill Scanlon, "Water Districts Sue EPA Over Two Forks Veto," *Rocky Mountain News,* November 23, 1991; US Department of Justice, "Judge Sides With US in Controversial Dam Project," press release, June 11, 1996.
69. James and Gerboth, "A Camp Divided," 163.
70. Ruth McManus and Philip J. Ethington, "Suburbs in Transition: New Approaches to Suburban History," *Urban History* 34, no. 2 (Aug. 2007): 324, 321; Martin V. Melosi, *The Sanitary City: Urban Infrastructure in America from Colonial Times to the Present* (Baltimore: Johns Hopkins Univ. Press, 2000), 355.
71. Hubert A. Farbes, *A New Path,* Policy Statement of the Denver Water Board (Denver: Board of Water Commissioners, April 17, 1992); Hubert A. Farbes, "Denver Can No Longer Promise to Supply Water to the Entire Metro Area," op-ed piece, *The Denver Post,* April 25, 1992; Hubert A. Farbes, "A New Path for the Denver Water Department," presentation to the 34th Annual Convention of the Colorado Water Congress, January 23, 1992; Ellison, *Denver Water Board,* 127–28. Farbes presented the New Path statement in several slightly different versions.
72. Farbes, "A New Path For the Denver Water Department."
73. Miller, *Water in the West,* 213; Luecke, "Two Forks," 69; Fisk, "Metro Denver Water Story," 129.
74. Arthur Conan Doyle, "The Adventure of Silver Blaze," *The Strand Magazine* (Dec. 1892); reprinted in *The Classic Illustrated Sherlock Holmes* (Stamford, CT: Longmeadow Press, 1987), 196–97.
75. Susan Consola Appleby, *Fading Past: The Story of Douglas County, Colorado* (Palmer Lake, Colorado: Filter Press, 2001), 33, 36, 38.
76. Jerry Brown, "Water Board Votes to Buy Town," *Rocky Mountain News,* December 23, 1981; Appleby, *Fading Past,* 42. See also Nell Fletcher and Nancy Hinshaw, *Echoes of Forgotten Places: The Places and their Echoes* (Mountain Artisans Arts Council), 2004.
77. Quoted in Appleby, *Fading Past,* 33.

Chapter Seven

Chipping Away at Tradition

1. Alex Prud'homme, *The Ripple Effect: The Fate of Freshwater in the Twenty-First Century* (New York: Scribner, 2011), 1.
2. Prud'homme, *The Ripple Effect,* 3–4.
3. Chips Barry, "'The World Would Be Different If Not for the Veto," interview by Ed Marston, *High Country News,* November 20, 2000.
4. Denise Lach, Helen Ingram, and Steve Rayner, "Maintaining the Status Quo: How Institutional Norms and Practices Create Conservative Water Organizations," *Texas Law Review* 83 (2005): 2028.
5. Lach, Ingram, and Rayner, "Maintaining the Status Quo," 2028–30.
6. Lach, Ingram, and Rayner, "Maintaining the Status Quo," 2032–34, 2038.
7. Lach, Ingram, and Rayner, "Maintaining the Status Quo," 2027, 2039, 2053.
8. Martin V. Melosi, *The Sanitary City: Urban Infrastructure in America from Colonial Times to the Present* (Baltimore: Johns Hopkins Univ. Press, 2000), 10.
9. Chips Barry, "Introductory Material" for a talk delivered April 9, 2008, Boulder, CO; Ed Marston, "Ripples Grow When a Dam Dies," *High Country News,* October 3, 1994, in Char Miller, ed., *Water in the West,* 206.
10. Chips Barry, personal interview with Patty Limerick, Jason L. Hanson, and Dylan Eiler, August 11, 2009, Denver Water, Denver, CO.

11. Chips and Gail Barry, "2009 Greetings from the Barry Family," letter to Patty Limerick; Barry, "Introductory Material."
12. "Memorial Program, Hamlet 'Chips' Barry III, 1944–2010," May 21, 2010.
13. Ben Wear, "Natural Resources Director Nominated," *Colorado Springs Gazette-Telegraph*, November 27, 1990; Mark Obmascik," "New Water Chief Bridges Ideologies: Barry Sees Self as Peacemaker," *The Denver Post*, December 3, 1990; Barry, personal interview.
14. Barry, personal interview.
15. Lach, Ingram, and Rayner, "Maintaining the Status Quo," 2033, 2039.
16. Denver Water, *Xeriscape Plant Guide: 100 Water-Wise Plants for Gardens and Landscapes* (Golden, CO: Fulcrum Publishing, 1996). See also David Winger, *Xeriscape Color Guide: 100 Water-Wise Plants for Gardens and Landscapes* (Golden, CO: Fulcrum Publishing, 1998).
17. J. Gordon Milliken, "Water Management Issues in the Denver Colorado Urban Area," in Mohamed T. El-Ashry and Diane C. Gibbons, *Water and Arid Lands of the Western United States* (New York: Cambridge Univ. Press, 1988), 333–75; Denver Water Planning and Finance Divisions, *Annual Report and Component Unit Financial Report for the Year Ended December 31, 1992* (Denver: Denver Water, 1993).
18. Mark Obmascik, "New Rates Aimed at Saving Water Board's Conservation Plan," *The Denver Post*, September 20, 1989; Chips Barry, "Continue Wise Use of Water," *The Denver Post*, September 9, 2005.
19. Rachel Brand, "Denver Water Resorts to Dry Humor: Lighthearted Ads Urge Customers to Conserve," *Rocky Mountain News*, July 2, 2002; Julie Dunn, "Denver Water's Ads Already Working Conservation Angle," *The Denver Post*, July 13, 2006; "Denver Water's Conservation Plan," www.denverwater.org/Conservation/Conservation Plan/, accessed August 4, 2011; Denver Water, *Solutions: Saving Water for the Future* (2010): 16.
20. Barry, personal interview; Integrated Resource Plan (1997), 4, 9, 42, 47.
21. Integrated Resource Plan (1997), 7.
22. Barry, personal interview; Chapter 5: Service Outside Denver—5.02 Distributor Contracts (Water Service Agreements), "Denver Water's Operating Rules" (as of 2011), available at www.denverwater.org/OperatingRules/OperRules5/; Integrated Resource Plan (1997),1–2.
23. Integrated Resource Plan (1997), 67.
24. US Census Bureau.
25. Aesop's Fables have appeared in numerous translations over the years. These quotations come "The Ant and the Grasshopper" at www.aesopfables.com.
26. Chips Barry, "Water Wrangler," interview by Natasha Gardner, *5280 Magazine* (Sept. 2008).
27. Mark Shively, personal communication, September 12, 2011 (Shively is the President of the Douglas County Water Resource Authority); Michael E. Long, "Colorado's Front Range," *National Geographic* 190, no. 5 (November 1996): 80–103; Bruce Finley, "Proposed Water Deal Aims to Creatively Ease Water Woes in Denver's South Suburbs," *The Denver Post*, Oct. 5, 2011.
28. Barry, personal interview.
29. Lach, Ingram, and Rayner, "Maintaining the Status Quo," 2039.
30. Barry, "Water Wrangler."
31. Barry, "Water Wrangler."
32. Lach, Ingram, and Rayner, "Maintaining the Status Quo," 2032; Integrated Resource Plan (1997), 6, 66;
33. Daniel F. Luecke, John Morris, Leo Rozaklis, and Robert Weaver, *What the Current Drought Means for the Future of Water Management in Colorado* (Sustainable Water Caucus, January 2003), 37, 46; "Reservoir May Triple in Size," (Boulder) *Daily Camera*, November 30, 2009; "Conservation Is Not Enough," *The Denver Post*, December

24, 2009; "County: Gross Reservoir Study 'inadequate,'" (Boulder) *Daily Camera,* March 21, 2010.

34. "Comments of The Environmental Group of Coal Creek Canyon, Coal Creek Canyon Improvement Association, Friends of the Foothills, and concerned citizens of the Gross Reservoir Community on Draft Environmental Impact Statement for the Moffat Collection System Project," March 17, 2010.
35. "Denver Water's current Moffat Tunnel diversions take, on average, over 50 percent of the native stream flows of the Fraser River," Trout Unlimited's Comments—Moffat Collection System Project, March 17, 2010, 5.
36. Kit Coddington, "Gross Reservoir: Environmental Impact Too Great," (Boulder) *Daily Camera,* March 13, 2010; Drew Peternell, "How Much Is a River Worth?" *The Denver Post,* July 5, 2011; "Permit Application NOW-2002080762-DEN comments Colorado Environmental Coalition," March 17, 2010, Comment Report (Organizations/Stakeholders Part B), 9–20; "Trout Unlimited's Comments—Moffat Collection System Project," March 17, 2010, Comment Report (Organizations/Stakeholders Part B), 44–86; "Western Resource Advocates Moffat DEIS Comments," March 17, 2010, Comment Report (Organizations/Stakeholders Part A), 169–206; and "Joint Rebuttal Report Regarding the Moffat Collection System Project DEIS," March 17, 2010, Comment Report (Jurisdictions/Municipalities Part B), 1–101.
37. Western Resource Advocates comments, 192.
38. Bruce Finley, "Dividing Line over Diverting Water," *The Denver Post,* December 9, 2009; Vincent Carroll, "How Little Water Is Enough?" *The Denver Post,* December 20, 2009.
39. "Executive Summary," Comments of The Environmental Group of Coal Creek Canyon et al, 3.
40. Board Resource Statement, October 15, 1996, Item IV-B; Integrated Resource Plan (1997), 65; Strategic Plan, "Organizational Perspective: Desired Outcome," fifth goal in the list.
41. "The Colorado River Cooperative Agreement," briefing summary, April 28, 2011.
42. Allan Best "Western Slope and Denver Seal Big Water Pact," *Mountain Town News,* April 18, 2011; "The Colorado River Cooperative Agreement," briefing summary, 3. See also "Colorado River Cooperative Agreement," April 28, 2011, Proposed Agreement, 46 (Definitions).
43. "The Colorado River Cooperative Agreement," briefing summary, 2.
44. Best, "Western Slope and Denver Seal Big Water Pact," April 18, 2011; "Colorado River Cooperative Agreement," April 28, 2011.
45. Drew Peternell, "How Much Is a River Worth?" *The Denver Post,* July 5, 2011; for Trout Unlimited's concerns, see, for example, "Trout Unlimited's Proposed Mitigations Measures for Inclusion in the Wildlife Commission's Recommended Mitigation Plan, Moffat Collection System Project," Wildlife Commission Meeting, May 6, 2011.
46. Integrated Resource Plan (1997), 6.
47. Best, "Western Slope and Denver Seal Big Water Pact," 2.
48. Colorado River Cooperative Agreement, Proposed Agreement, April 28, 2011, p. 9, 3, 7, and 10.
49. Barry, personal interview.

Conclusion

Turning Hindsight into Foresight

1. Thomas Hornsby Ferril, *Thomas Hornsby Ferril and the American West,* eds., Robert C. Baron, Stephen J. Leonard, and Thomas J. Noel (Golden, CO: Fulcrum Publishing and the Center of the American West, 1996), 36.
2. John Kenneth Galbraith, *The Affluent Society,* Fortieth Anniversary Edition (Boston: Houghton Mifflin, 1998), 11.

3. Denver Water, "Key Facts," available at DenverWater.org.
4. Western Resource Advocates Comments, 190 (or 17), 195 (or 22).
5. Elizabeth Kohlbert, "Turf War," *New Yorker* (July 21, 2008), 80–86; Ted Steinberg, *American Green: The Obsessive Quest for the Perfect Lawn* (New York: WW Norton, 2006), 204.
6. Quoted in Steve Hinchman, "EPA to Denver: Wake Up and Smell the Coffee," reprinted in Miller, *Water in the West,* 206.
7. Walter Eha, *Water for Denver: How Water Helped Build a City* (Denver: Denver Water/ Denver Public Library, 2006), 10, 29; Merson quoted in Suzanne Weiss, "Regional EPA Chief Urges Foothills' Demise," *The Denver Post,* September 27, 1977; Brian A. Ellison, *The Denver Water Board: Bureaucratic Power and Autonomy in Local Natural Resource Agencies* (PhD dissertation, Colorado State Univ.–Ft. Collins, 1993), 29, 33–34.
8. J. Gordon Milliken, "Water Management Issues in the Denver, Colorado, Urban Area," in *Water and Arid Lands of the Western United States,* eds., Mohamed T. El-Ashry and Diana C. Gibbons (Cambridge Univ. Press, 1988), 366–67.
9. Milliken, "Water Management Issues," 348; Daniel F. Luecke, personal interview with Patty Limerick, Jason L. Hanson, and Dylan Eiler, October 19, 2009, Center of the American West at the University of Colorado at Boulder; Hal Rothman, "Water and the Future of Las Vegas," reprinted in Miller, *Water in the West,* 242.
10. *Charter of the City and County of Denver, Adopted by a Vote of the People May 19, 1959,* Article X, Sections 10.1.13, 10.1.9, 10.1.10.
11. Charles Howe, "Alternatives Needed from Water Board" (letter to the editor), *The Denver Post,* July 8, 1972.
12. Donald Worster, *Rivers of Empire: Water, Aridity, and the Growth of the American West* (New York: Pantheon Books, 1985), 7.
13. Donald J. Pisani, *To Reclaim a Divided West: Water, Law, and Public Policy, 1848–1902* (Albuquerque: Univ. of New Mexico Press, 1992), xvi–xvii.
14. Norris Hundley Jr., *The Great Thirst: Californians and Water, 1770s–1990s* (Berkeley: Univ. of California Press, 1992), xv, xvi.
15. Steven P. Erie, *Beyond Chinatown: The Metropolitan Water District, Growth, and the Environment in Southern California* (Palo Alto, CA: Stanford Univ. Press, 2006), 14.
16. Carl Abbott, *How Cities Won the West: Four Centuries of Urban Change in Western North America* (Albuquerque: Univ. of New Mexico Press, 2008), 154.
17. Thomas Jefferson, *Notes on the State of Virginia,* ed. Thomas Perkins Abernethy (New York: Harper Collins, 1964), 157–58.
18. Colorado Water Conservation Board, "Section 4-3: Agricultural Consumptive Needs" and "Mission Statement, Key Findings, and Recommendations," *Statewide Water Supply Initiative—2010* (Denver, Colorado Water Conservation Board, 2010); Center of the American West, *Atlas of the New West,* William E. Riebsame et al., eds. (New York: W.W. Norton, 1997), 82–84. The Colorado Water Conservation Board calculates that Colorado agriculture's share of the state's water use will decline from 86 percent today to 82 percent in 2050. Roughly 86 percent of the South Platte River goes to agricultural uses on the plains east of Denver, according to Scott Willoughby, "Wonder Waters: The South Platte," *The Denver Post,* July 8, 2007.
19. Kenneth Jackson, *Crabgrass Frontier,* 288.
20. David Owen, *Green Metropolis: Why Living Smaller, Living Closer, and Driving Less Are the Keys to Sustainability* (New York: Riverhead Books, 2009), 8–10, 25, and 27. See also Edward Glaeser, *Triumph of the City: How Our Greatest Invention Makes Us Richer, Smarter, Greener, Healthier, and Happier* (New York: Penguin Press, 2011).
21. Martin V. Melosi, *The Sanitary City: Urban Infrastructure in America from Colonial Times to the Present* (Baltimore: Johns Hopkins Univ. Press, 2000), 4; Christopher P. Leinberger, "The Death of the Fringe Suburb," *The New York Times,* November 26, 2011.

22. Western Resource Advocates, Trout Unlimited, and Colorado Environmental Coalition, "Filling the Gap: Commonsense Solutions for Meeting Front Range Water Needs" (2011).
23. William Cronon, *Nature's Metropolis: Chicago and the Great West* (New York: W.W. Norton, 1991), 7, 384.
24. Melosi, *Sanitary City*, 294; Diane Raines Ward, *Water Wars: Drought, Flood, Folly, and the Politics of Thirst* (New York: Penguin Putnam Inc., 2002), 201.
25. Robert Glennon, *Unquenchable: America's Water Crisis and What to Do about It* (Washington, DC: Island Press, 2009), 105.
26. Guy Rocha, "Myth #122—What Mark Twain Didn't Say," Nevada State Library and Archives, online at http://nsla.nevadaculture.org/index.php?option=com_content&task=view&id=803&Itemid=418.
27. Wendy Barnaby, "Do Nations Go to War Over Water?" *Nature* 458 (March 19, 2009): 282–83; Tony Allan, *Virtual Water: Tackling the Threat to Our Planet's Most Precious Resource* (London: I.B. Tauris, 2011).
28. Alexander Bell, *Peak Water: Civilisation and the World's Water Crisis* (Edinburgh: Luath Press, 2009), 182.
29. Denver Water Board, "Customers Served" and "Denver Water Board's Water Use," in "Key Facts," available online at www.denverwater.org/AboutUs/KeyFacts; Colorado Water Conservation Board, *Statewide Water Supply Initiative—2010* (Denver: Colorado Water Conservation Board, 2010); Gregory J. Hobbs Jr., *Citizen's Guide to Colorado Water Law* (Denver: Colorado Foundation for Water Education, 2004), 7.
30. Barnaby, "Do Nations Go to War Over Water?" 283.

Further Readings

Denver Water History

Bacow, Lawrence S., and Michael Wheeler. "Mediating Large Disputes." In *Environmental Dispute Resolution.* New York: Plenum Press, 1984.

Carpenter, Susan L., and W. J. D. Kennedy. "The Denver Metropolitan Water Roundtable: A Case Study in Researching Agreements." *Natural Resources Journal* 28 (Winter 1988): 21–35.

Cox, James L. "The Development and Administration of Water Supply Programs in the Denver Metropolitan Area." PhD diss., Univ. of Colorado–Boulder, 1965.

———. *Metropolitan Water Supply: The Denver Experience.* Boulder, CO: Bureau of Governmental Research and Service and the Univ. of Colorado, 1967.

Eha, Walter. *Water for Denver: How Water Helped Build a City.* Denver: Denver Water/Denver Public Library, 2006.

Ellison, Brian A. "Autonomy in Action: Bureaucratic Power and Autonomy in Local Natural Resource Agencies." *Policy Studies Review* 14, no. 1/2 (Spring/Summer 1995): 25–48.

———. "Denver Water Politics, Two Forks, and the Implication for Development on the Great Plains." In *Water on the Great Plains: Issues and Policies,* edited by Peter J. Longo and David W. Yoskowitz, 93–115. Lubbock, TX: Tech Univ. Press, 2002.

———. "The Denver Water Board: Bureaucratic Power and Autonomy in Local Natural Resource Agencies." PhD diss., Colorado State Univ.–Ft. Collins, 1993.

Fisk, Charles C. "The Metro Denver Water Story: A Memoir." Unpublished manuscript, (Fort Collins: Colorado State Univ. Libraries, 2005).

Hobbs Jr., Gregory J. "Green Mountain Reservoir: Lock or Key?" *Colorado Water Rights* 3, no. 2 (Summer 1984): 1–11.

———. "State Water Politics Versus an Independent Judiciary: The Colorado and Idaho Experiences." *Quinnipiac Law Review* 20 (2001): 669–96.

King, Clyde. *The History of the Government of Denver with Special Reference to its Relations with Public Service Corporations.* Denver: Fisher Book Company, 1911.

Luecke, Daniel F. "Two Forks Dam and Endangered Species." *Colorado Water* 26, no. 2 (March/April 2009): 17–19.

———."Two Forks: The Rise and Fall of a Dam." *Natural Resources and Environment* 14 (Summer 1999): 24–28, 69.

Milliken, J. Gordon. "Water Management Issues in the Denver, Colorado, Urban Area." In *Water and Arid Lands of the Western United States,* edited by Mohamed T. El-Ashry and Diana C. Gibbons. 333–76. Cambridge: Cambridge Univ. Press, 1988.

Mosley, Earl L. *History of the Denver Water System, 1858–1919.* Denver: Board of Water Commissioners, 1962.

———. "Western Slope Water Development for Denver." *Journal of the American Water Works Association* 49 (March 1957): 251–62.

Perkin, Robert. *The First Hundred Years: An Informal History of Denver and the Rocky Mountain News*. Garden City, NY: Doubleday & Company, 1959.
Shoumatoff, Alex. "The Skipper and the Dam." *The New Yorker*, December 1, 1986.
Smiley, Jerome C. *History of Denver: With Outlines of the Earlier History of the Rocky Mountain Country*. Denver: The Times–Sun Publishing Company, 1901.
Tyler, Daniel. *The Last Water Hole in the West: The Colorado-Big Thompson Project and the Northern Colorado Water Conservancy District*. Niwot: Univ. Press of Colorado, 1992.

Denver and Colorado History

Abbott, Carl, Stephen Leonard, and Thomas Noel. *Colorado: A History of the Centennial State*. Niwot: Univ. Press of Colorado, 2005.
Arps, Louisa Ward. *Denver in Slices: A Historical Guide to the City*. Denver: Sage Books, 1959.
Berwanger, Eugene. *The Rise of the Centennial State: Colorado Territory, 1861–1876*. Urbana: Univ. of Illinois Press, 2007.
Brosnan, Kathleen A. *Uniting Mountain and Plain: Cities, Law, and Environmental Change along the Front Range*. Albuquerque: Univ. of New Mexico Press, 2002.
Brown, F. Lee and Helen M. Ingram. *Water and Poverty in the Southwest*. Tucson: Univ. of Arizona Press, 1987.
Dorsett, Lyle. *The Queen City: A History of Denver*. Boulder, CO: Pruett Publishing Company, 1977.
Hafen, Leroy R. *Colorado: A Story of the State and Its People*. Denver: The Old West Publishing Company, 1948.
Haigh, Jane G. "Interests and the Political Economy of Corruption." In *Political Power, Patronage, and Protection Rackets: Con Men and Political Corruption in Denver, 1889–1894*. PhD diss., Univ. of Arizona, 2009.
Hobbs, Jr., Gregory J. *Citizen's Guide to Colorado Water Law*. Denver: Colorado Foundation for Water Education, 2003.
James, Franklin J., and Christopher B. Gerboth. "A Camp Divided: Annexation Battles, the Poundstone Amendment, and Their Impact on Metropolitan Denver, 1941–1988." *Colorado History* 5 (2001): 129–74.
Leonard, Stephen and Thomas Noel. *Denver: Mining Camp to Metropolis*. Niwot: Univ. Press of Colorado, 1990.
Romero II., Tom. *Of Race and Rights: Legal Culture, Social Change, and the Making of a Multiracial Metropolis, Denver, 1940–1975*. PhD diss., Univ. of Michigan, 2004.
West, Elliott. *The Contested Plains: Indians, Goldseekers, and the Rush to Colorado*. Lawrence: Univ. Press of Kansas, 1998.

Western American and Water History

Abbott, Carl. *How Cities Won the West: Four Centuries of Urban Change in Western North America*. Albuquerque: Univ. of New Mexico Press, 2008.
Adler, Robert. *Restoring Colorado River Ecosystems: A Troubled Sense of Immensity*. Washington DC: Island Press, 2007.
Center of the American West, *Atlas of the New West*, William E. Riebsame et al., eds. New York: W. W. Norton, 1997.

Cronon, William. *Nature's Metropolis: Chicago and the Great West.* New York: W. W. Norton, 1991.

deBuys, William. *A Great Aridness: Climate Change and the Future of the American Southwest.* New York: Oxford Univ. Press, 2011.

deBuys, William, and Joan Myers. *Salt Dreams: Land and Water in Low-Down California.* Albuquerque: Univ. of New Mexico Press, 1999.

Elkind, Sarah. *Bay Cities and Water Politics: The Battle for Resources in Boston and Oakland* Lawrence: Univ. Press of Kansas, 1998.

Erie, Steven P. *Beyond Chinatown: The Metropolitan Water District, Growth, and the Environment in Southern California.* Stanford: Stanford Univ. Press, 2006.

Fiege, Mark. *Irrigated Eden: The Making of an Agricultural Landscape in the American West.* Seattle: Univ. of Washington Press, 1999.

Harvey, Mark. *A Symbol of Wilderness: Echo Park and the American Conservation Movement,* Weyerhaeuser Environmental Classic ed. Seattle: Univ. of Washington Press, 2000.

Hundley, Norris, Jr. *The Great Thirst: Californians and Water, 1770s–1990s.* Berkeley: Univ. of California Press, 1992.

———. *Water and the West: The Colorado River Compact and the Politics of Water in the American West.* Berkeley: Univ. of California Press, 1975.

Kahrl, William L. *Water and Power: The Conflict over Los Angeles' Water Supply in the Owens Valley.* Berkeley: Univ. of California Press, 1982.

Klingle, Matthew. *Emerald City: An Environmental History of Seattle.* New Haven, C: Yale Univ. Press, 2007.

Koeppel, Gerald T. *Water for Gotham.* Princeton: Princeton Univ. Press, 2001.

Layton, Edwin Jr. *The Revolt of the Engineers: Social Responsibility and the American Engineering Profession.* Cleveland: The Press of Case Western Reserve Univ., 1971.

Logan, Michael. *Desert Cities: The Environmental History of Phoenix and Tucson.* Pittsburgh: Univ. of Pittsburgh Press, 2006.

Miller, Char, ed. *Water in the West: A High Country News Reader.* Corvallis: Oregon State Univ. Press, 2000.

Pisani, Donald J. *To Reclaim a Divided West: Water, Law, and Public Policy, 1848–1902.* Albuquerque: Univ. of New Mexico Press, 1992.

———. *Water and the American Government: The Reclamation Bureau, National Water Policy, and the West, 1902–1935.* Berkeley: Univ. of California Press, 2002.

Reisner, Marc. *Cadillac Desert: The American West and Its Disappearing Water.* New York: Viking, 1986.

Reuss, Martin and Stephen H. Cutcliffe. *The Illusory Boundary: Environment and Technology in History.* Charlottesville: Univ. of Virginia Press, 2010.

Righter, Robert W. *The Battle over Hetch Hetchy: America's Most Controversial Dam and the Birth of Modern Environmentalism.* New York: Oxford Univ. Press, 2005.

Rowley, William D. *The Bureau of Reclamation: Origins and Growth to 1945.* Washington, DC: US Dept. of the Interior, 2006.

Rodriquez, Sylvia. *Acequia: Water Sharing, Sanctity, and Place.* Santa Fe: School for Advanced Research Press, 2006.

Shurts, John. *Indian Reserved Water Rights.* Norman: Univ. of Oklahoma Press, 2000.

Wiley, Peter, and Robert Gottlieb. *Empires in the Sun: The Rise of the New American West.* Tucson: Univ. of Arizona Press, 1985.

Wilkinson, Charles. *Crossing the Next Meridian: Land, Water, and the Future of the West.* Washington, DC: Island Press, 1992.

Worster, Donald. *Rivers of Empire: Water, Aridity, and the Growth of the American West.* New York: Pantheon, 1985.

National and International Context

Allan, Tony. *Virtual Water: Tackling the Threat to our Planet's Most Precious Resource.* London: I.B. Tauris, 2011.

Barnaby, Wendy. "Do Nations Go to War Over Water?" *Nature* (March 19, 2009): 282–83.

Bell, Alexander. *Peak Water: Civilisation and the World's Water Crisis.* Edinburgh: Luath Press, 2009.

Glaeser, Edward. *Triumph of the City: How Our Greatest Invention Makes Us Richer, Smarter, Greener, Healthier, and Happier.* New York: Penguin Press, 2011.

Glennon, Robert. *Unquenchable: America's Water Crisis and What to Do About It.* Washington, DC: Island Press, 2009.

Lach, Denise, Helen Ingram, and Steve Rayner. "Maintaining the Status Quo: How Institutional Norms and Practices Create Conservative Water Organizations." *Texas Law Review* 83 (2005): 2027–53.

MacDonnell, Lawrence. *From Reclamation to Sustainability: Water, Agriculture, and the Environment in the American West.* Niwot: Univ. of Colorado Press, 1999.

Melosi, Martin. *Precious Commodity: Providing Water for America's Cities.* Pittsburgh: Univ. of Pittsburgh Press, 2011.

Melosi, Martin. *The Sanitary City: Urban Infrastructure in America from Colonial Times to the Present.* Baltimore: John Hopkins Univ. Press, 2000.

Owen, David. *Green Metropolis: Why Living Smaller, Living Closer, and Driving Less Are the Keys to Sustainability.* New York: Riverhead Books, 2009.

Pollan, Michael. "Why Mow?" In *Second Nature: A Gardner's Education.* New York: Grove Press, 1991.

Pritchard, Sara B. *Confluence: The Nature of Technology and the Remaking of the Rhone.* Cambridge, MA: Harvard Univ. Press, 2011.

Prud'homme, Alex. *The Ripple Effect: The Fate of Freshwater in the Twenty-First Century.* New York: Scribner, 2011.

Pursell, Carroll. *The Machine in America: A Social History of Technology.* 2nd ed. Baltimore: Johns Hopkins Univ. Press, 2007.

Rawson, Michael. *Eden on the Charles: The Making of Boston.* Cambridge, MA: Harvard Univ. Press, 2010.

Solomon, Steven. *Water: The Epic Struggle for Wealth, Power, and Civilization.* New York: Harper Collins, 2010.

Ted Steinberg, *American Green: The Obsessive Quest for the Perfect Lawn.* New York: W. W. Norton, 2006.

Ward, Diane Raines. *Water Wars: Drought, Folly, and the Politics of Thirst.* New York: Riverhead/Penguin Putnam, 2002.

White, Richard. "Information, Markets, and Corruption: Transcontinental Railroads in the Gilded Age." *Journal of American History* 90, no. 1 (June 2003): 19–43.

Creative Nonfiction About Water

Abbey, Edward. *Desert Solitaire: A Season in the Wilderness.* New York: McGraw-Hill, 1968.

Childs, Craig. *The Secret Knowledge of Water.* New York: Back Bay Books, 2000.

Crawford, Stanley. *Mayordomo.* Albuquerque: Univ. of New Mexico Press, 1988.

Index

M

N

O

P

About the Author

Patricia Nelson Limerick is the faculty director and board chair of the Center of the American West at Colorado University, where she is also a professor of history and environmental studies. She has received a MacArthur Fellowship and a number of other awards and honors. She currently serves as the vice president for the Teaching Division of the American Historical Association. Her most widely read book, *The Legacy of Conquest,* is in its twenty-fifth year of publication.